Modern Wood Finishi

MODERN Wood Finishing Techniques

Noel Johnson Leach

Fully Illustrated

STOBART DAVIES
HERTFORD

Copyright © 1993 Noel Johnson Leach

All rights reserved. No part of this publication may be reproduced, stored in a retrieval system, or transmitted, in any form or by any means, electronic, mechanical, photocopying, recording or otherwise, without the prior permission of the copyright owner or the publishers.

British Library Cataloguing in Publication Data
A catalogue record for this book is available from the British Library.

ISBN 0–85442–056–8

Published 1993 by Stobart Davies Ltd, Priory House, Priory Street, Hertford SG14 1RN.
Set in 10 on 13.5 pt Bembo by Ann Buchan (Typesetters) Shepperton.

ACKNOWLEDGEMENTS

The author wishes to convey his sincere gratitude to the following companies and individuals for supplying various photographs and their help in providing information on new equipment and materials.

Mr R. Rustin of Rustins Ltd
DeVilbiss Ransburg Company (UK)
Sonneborne & Rieck Ltd
Kremlin spray Painting Equipment Ltd
Magnum Compressors Ltd
Gibbs Finishing systems Ltd
John Myland Ltd
P.J. Dust Extraction Ltd
The Hydrovane Compressor Co.Ltd
Rentokil Ltd
W.S. Jenkins & Co.Ltd
Richard Sorsky

(Full addresses of companies are given at the back of the book)

I am also indebted to my wife, Joan, for her invaluable help in correcting my scripts and re-typing the text in a presentable form, and for her help in the photographic sessions. Also my thanks to my daughter, Helen, who helped with the typing; and to my daughter, Mrs N.D. Mularczyk, SRN, SCM, NDN, for the help with the medical section and update on skin complaints in the Health and Safety chapter.

CONTENTS

		page
1.	Introduction	9
2.	Preparation of Wood Substrates	13
3.	Abrasives	19
4.	Wood Infestation	31
5.	Bleaching	38
6.	Stripping and Cleaning Wood Surfaces	46
7.	Oils, Waxes, Lubricants, Solvents, Thinners	55
8.	Metal Finishing	66
9.	Colouring Wood	73
10.	Polishing Wood on the Lathe	90
11.	Shellac and French Polishing	96
12.	Distressing and Other Exotic and Simulated Finishes	109
13.	Temperamental Substrates	128
14.	Modern Wood Finishing Lacquers	135
15.	Spray Finishing	155
16.	Spray Booths & Equipment	188
17.	Health and Safety	196
18.	Questions and Answers	207
19.	List of Suppliers	223
	Index	228

CHAPTER ONE

Introduction

Wood finishing has been practiced for thousands of years, not just for utilitarian use but also for the decorative purposes. The ancient Egyptians and Persians in particular brought this decorative aspect to a fine art by applying gold leaf to their furniture and tomb artefacts. Indeed, gold leaf is still used to this day as, for example, in the refurbishment of the House of Lords some years ago.

The first recorded woodfinisher was Noah! In Genesis, Ch 6, v 14, God gave Noah working instructions 'Make thee an ark of gopher wood; rooms shalt thou make in the ark, and shalt pitch it within and without with pitch'. The biblical wood 'gopher' could have been either cypress or cedar, both very durable resinous woods. Pitch has been used for most marine woodfinishing down the ages and is still used to this day in some form or another. The Romans, those great woodfinishers, used vast amounts of pitch or tar for the preservation of their weapons, boats and buildings. As well as pitch, 'charing' was used on all exterior surfaces where the wooden posts, etc, were placed into the ground as this prevented decay; it was a method of burning the wood to a depth of 6mm/¼ inch or more so that it would not support fungal growth.

Down the centuries, Man has preserved his boats with varnishes in which the Greeks specialised. Vegetable oils, resins from trees such as gum arabic from the acacia tree, turpentine from the long leaf pines, lac resins from the secretion of the *Laccifer lacca* scale insect, resulting, in more modern times, to 'shellac', and beeswax from the female bee. All these have helped provide ready-made natural products for protecting wood against rot and deterioration, and to improve the standard of living brought about by the use of these products, long before the industrial chemists arrived on the scene.

Throughout the 18th century lac was a crude finish, until the 19th century saw the development of 'shellac'. During the mid-1800's the finish for wood and furniture was largely either varnish or french polish. By the middle of the 1900's the finish for most furniture was cellulose based surface coatings, and now, nearing the end of the century, water-borne surface coatings (acrylics) have been developed and are the coating for furniture and most other wooden surfaces – not in any way replacing solvent finishes so far, but a major competitor.

The words 'wood finishing' to the average person conjures up a mental picture of a very old gentleman with small-rimmed glasses rubbing away at a piece of furniture, or some young thing wielding a new paint brush charged from a tin of the latest varnish or paint just on the market. This is the TV commercial image of the trade in general. Not withstanding these images the wood finisher today has still to suffer the 'Do you actually make a living from it?' type of question.

Wood finishing, like anything else in life today, has changed out of all recognition to what it was 'just yesterday'. New products are coming on to the market almost daily; new materials, new appliances, machines etc, and new processes are prominent in the trade and popular press. Solvent-free products now being produced are the materials of the future. Who would have thought a few years ago that on the market now and in common use would be water-based surface coatings, such as gloss paints, lacquers and varnishes? The 'in' word is 'acrylic', and it is here to stay, now that these water-based finishes are in common use, because the public, in its obsession for environmentally-friendly products, is demanding new non-chemical, solvent-free surface coatings.

Through these pages I hope to introduce you to these new products and methods of application and to lay aside the old out-moded mysteries surrounding woodfinishing. Do not, for one moment, think that I have forsaken the well-tried and proven traditional methods, but during the many years that I have been a wood finisher, if I have found a 'better' less time-consuming method which still produces a first class result, I would use it and develop it.

We are all living and enjoying the technology of the latter part of the 20th century and some of it has been introduced into wood finishing in the preparation and in the many finishes that protect our most vital natural commodity – wood.

The object of this book is to collate the experiences of my work as a wood finisher, and with the knowledge accumulated over the past 40 years as a practical woodfinisher, lecturer and writer, instill, help and guide through the complex nature of this craft all who may require help. They may be furniture makers, bedroom and kitchen fitters and makers, journeymen, hotel maintenance staff, candidates for City & Guilds examinations, managers of paint stores, builders, architects, and last but not least, the D.I.Y. enthusiasts – the list is endless.

Today the woodfinisher is confronted with an ever-increasing array of new materials and equipment which can lead to confusion. This fact is obvious, for even from various commercial suppliers of wood finishing materials, their catalogues are riddled with ambiguity and out-dated instructions. Wood finishers on the whole are mostly a conservative lot, and for many decades the trade has been shrouded with mystery, ignorance and the tendency to ignore any new method or material that finds its way on to the trade market. Yet slowly, mainly due to commercial necessity and increasing awareness by the general public of the use of environmentally-friendly products in the form of water-based surface coatings, these have become the vogue and style of future woodfinishing. This book is aimed at practical exponents of the craft, but with the reservation that before anyone becomes involved with any method of woodfinishing, they must have at least some understanding of the make up of the materials available to them, and to the best way they can apply these finishes successfully.

Woodfinishing can be classed as an art form. Indeed, when it comes to exotic finishes and methods of application, a wood finisher must study the nature of wood and try and understand it in just the same way as an artist painter may study nature. This is particularly so in respect of the new species of woods that during the past few years have appeared on the market, and he or she must learn how these respond to various treatments such as staining, bleaching, oiling, painting, polishing and spraying.

The title of the book 'Modern Wood Finishing Techniques' may perplex the reader who perhaps wonders what is to follow. Unlike some instructions of high technology products that assumes the user knows how to operate or use the product and leaves out essential working instructions, this

Introduction

book makes no such assumption or omissions; and nor do I assume that the reader knows the meaning of any abbreviation used – I always give the full meanings. In each chapter I outline the traditional methods and materials used to produce a finish followed by any up-to-the-minute improvements which still produce the same high standard finish.

Basically, traditional methods use organic or even a mixture of organic and synthetic products simply to speed up the process of drying, while chemically produced products known as 'modern finishings' must be kept apart and used completely differently in application, these methods will be fully explained within the relevant chapters.

The subject can be fascinating, and a whole new world open up for the user of modern finishing methods; with new materials and the use of a spray gun, there is a whole host of processes and finishes not possible to achieve by traditional methods.

I have included a chapter on metal finishing and treating corrosion which may surprise some as the book is predominantly on wood finishing, but most furniture and joinery has some metal component such as hinges, locks, catches, handles, bands and stays, whilst some modern furniture has a percentage of metal used in the design construction.

In the forthcoming pages I have generally mentioned products that are easily obtained through the normal suppliers and outlets, although one has to be wary of products that disappear from the market shelves quite quickly, or that change their names, trade names, or even company names.

At the end of most chapters I have included a Health and Safety check list as a guide to the user, which I think is an advantage over leaving such information until the end of the book, and it is intended as a quick reference when using some of the modern chemical solvent products. There is, of course, a full Health and Safety chapter included in the book, coupled with references to UK COSSH Regulations (Control of Substances Hazardous to Health) and the US OSHA regulations (Occupational Safety & Health Administration).

I hope that as the pages unfold the newcomers and students to the trade or craft, and even some of the older craftsmen, may learn something to their advantage, and if this is achieved I shall be more than happy.

No matter what method of wood finishing is used and no matter what equipment is purchased, the ultimate quality of the finish will show up in the ability of the operator to produce a quality finish using one item which time will never replace – craftsmanship.

Note
Throughout the text references are made to items which may differ in terminology or name to those used in North America or other parts of the world. In most cases equivalent manufactured products are available from finishing suppliers, trade houses and stores.

A few of the differences are noted here, others are bracketed within the text.

UK Term	*USA Term*
.880 Ammonia	= Aqueous ammonia or 26% industrial ammonia
Caustic soda	= Lye
China clay	= Kaolin

Cissing	= Fish eye
Cellulose thinners	= Lacquer thinners
Finishing spirit	= Finishing or clear alcohol
Lubrisil (trade name)	= Silicon carbide abrasive paper (self lubricating)
Methylated spirits	= Alcohol
Mutton cloth	= Pure cotton mesh cloth
Paraffin	= Kerosene (lamp oil)
White spirit/turp substitute	= Mineral spirit

CHAPTER TWO

Preparation of Wood Substrates

No matter what equipment is being used, or however sophisticated, the quality of finish depends upon the preparation of the surface before any kind of surface coating can be applied, whether it is varnish, oil or wax, lacquer, paint or acrylic surface coatings. The preparation of the substrate must be as smooth and clean a surface as possible to enable the wood finisher to apply stains, fillers, sealers and finishing coats. Ignore the process of good preparation and expect poor quality surface coating films: spend time and effort on the preparation and the finished result will be as near perfect as the finisher is able to produce.

Wood comes to the consumer in various forms – in its natural solid state, in veneers and in many man-made products such as chipboard, blockboard, plywood and MDF (Medium Density Fibreboard), all of which require their own method of preparation and, in some of the cases, special attention to the finishing techniques.

First of all examine the surface and pin-point the faults: these could be bad planing or saw marking or other machine made faults, holes, splits, knots, etc. All must be treated and made good so that they do not show up in the finishing. To remove such surface faults the use of either a cabinet scraper or sanding machine is ideal, followed finally by hand sanding in the direction of the grain, and it is here that attention to sanding is paramount. The choice of abrasive papers should be either 150–240 grade aluminium oxide or garnet, depending upon the hardness of the wood involved.

Then there is the decision as to whether or not to fill the grain. The filling of grain is becoming less popular due to the commercial cuts in production line time and also due to the present vogue in finishes. In the 1930's wood was finished to look like glass, but not so nowadays. Wood 'texture' has become more important to the full effect of the final finish which has been long overlooked in past decades.

The materials required for the preparation work are the following:

1. Abrasive coated sheets (280 × 230 mm) or abrasive belts, pads, etc, if using power assisted machines. The grades of sanding papers I would suggest are a selection of aluminium oxide 80 to 150 grades and garnet 240 for final finishing, but this all depends upon the wood. The above abrasives are ideal for hardwoods such as oak, chestnut, ash etc, but the softer fibres of woods such as lime, mahogany, etc, would benefit more by using garnet grades throughout.

2. Wood fillers

3. Wood grain fillers
4. Stoppers
5. Putty
6. Tools and other materials required are a cabinet scraper, sanding block, soldering iron, power sanding equipment, coarse hessian, veneer knife, adhesives.

Wood fillers

These must not be confused with wood grain fillers and are for use when you need to fill a large gash on a substrate yet not necessarily fill the grain. These fillers come in pastes ready for use – plastic wood is one example of a wood filler, whilst putty is another if you are treating a substrate for pre-painting. On the market today are many examples of catalysed wood fillers which dry within a few minutes and can then be sanded, painted or varnished. These are excellent modern wood fillers and will not 'sink' as old traditional wood fillers had a tendency to do. The more traditional word 'stoppers' is often used to describe the materials for filling small indentations on wood substrates during the french polishing technique. These stoppers are soft to hard coloured pigmented waxes specially prepared for the purpose of pushing into small crevices and rubbing flat to the contours of the surface. They are compatible with many finishes although great care must be taken if the surface coating is going to be an acid catalysed type. They must not be used for large holes as they have a tendency to sink and thus show up as a minor concave fault.

Another name for stopper is sometimes called 'beaumontage' which is made from beeswax, crushed rosin and shellac. These are softer in texture and they are supplied in many colours. Shellac sticks, however, are also used in the filling of holes and cracks etc, and these are also made in many colours but have a tendency to fall out of the crevices. They require to be melted first with a hot iron and then the hot wax is dropped onto the area required, and abrasive paper is then used to level it out. The only advantage of these shellac sticks is that they instantly harden on contact with the wood but will not sink.

Proprietary stoppers

There are three basic forms of 'stoppers', and great care must be taken in the correct choice when using these fillers. The correct form for use under shellac polishes is the cake-type stopping material which is a mixture of natural resins and lead-free pigments and must only be used under shellac or french polish finishes. It is supplied in many colours.

The cellulose cake stopping is for use under cellulose surface coating lacquers and pre-catalysed finishes only. They are a mixture of natural resins and lead-free pigments with no traces of animal waxes, such as beeswax, and are supplied in many colours such as teak, light and dark oak, sapele, mahogany, walnut, white, black and yew.

The third type which is specially made for use under *all* catalysed lacquers is called stick stopping, and is supplied in the normal colour range.

When using catalyst lacquers a reaction can occur if beeswax is used, causing the area containing the stopper not to dry. It is therefore important for the correct stopper to be used.

Another well-known traditional stopper (sometimes called cabinetmakers putty) is made from using the actual fine sawdust from woods such as mahogany, oak, etc, and mixing them with a little adhesive – either hot animal glue, cold PVA glue or 'Cascamite' glue. When dry they can be sanded down flat and no tinting of the filled area need be required. The only snag here is that on staining, the filled area will absorb stain at a different rate than the surrounding wood, but a little touch of shellac before staining on the filled area prevents this occurance.

Commercially manufactured ready-mixed paste stopper is available in various colours and is a slow-drying stopping compound which is easily sanded and is mainly used by cabinet makers and joiners.

Wood grain fillers

These fillers are intended for the filling of wood grain only. They are either a powder or paste which is spread onto the wood and rubbed across the grain using a piece of hessian. This is left on the wood for a little while and then rubbed hard across the grain using a dry piece of hessian, thus removing all traces from the actual surface and simply leaving the filler in the grain. The rubbing action across the grain prevents the grain filler from being removed from the grain crevices. The filler is then allowed to dry hard and then sanded to obtain a smooth finish. Another reason for filling grain is to prevent excessive absorption of stains and other finishing surface coatings which are now costly. Grain fillers come in various forms, but one particular grain filler will not be compatible with all surface coatings, so great care is required to make sure that the grain filler being used will not upset the balance of the surface coating when applied.

The make-up of a wood grain filler is complex: it consists of a filling powder, usually amorphous silica, while 'extenders' are added which are padding agents to body up the filler – usually fine clay or chalk, which also gives texture. A 'binder' is added which could be oil, resin or gold size – determined by the type of filler required – together with pigments or dyes. Finally a solvent, which is the means of uniting and spreading the compound, which could be white spirit, naphtha or other synthetic solvents.

There are basically four types of wood grain filler available:

1. *Water bound wood grain fillers*

These are supplied in two forms: the first in powder, the second in paste. In the powder type, the addition of water is all that is required, while the paste is ready mixed for use. Popular ready-mixed fillers are available from general D.I.Y. stores and they give a fine smooth finish to any kind of wood surface. They are ideal for all substrates that are to be painted with either oil solvent paints, emulsion or acrylic finishes. Whilst they are ideal for pigmented surface coatings they are not recommended for clear surface finishes as the filler in the grain will always show through the surface film as white in the grain. These fillers can also be used on plaster or fibreboard, etc.

Popular brands in the UK and USA are Polycell's Fine Surface Polyfiller, and Benwood Paste Wood Filler.

2. *Oil bound wood grain fillers*

These are popular traditional fillers and they are often called 'patent fillers' which use an oil as binder, and most trade houses have their own house brands. They are ideal for all surface coatings such as french polish, shellac varnishes, oil varnishes (although a coat of shellac sealer is recommended before coating oil varnishes simply to prevent bleeding). These oil bound fillers require twelve to twenty-four hours to dry hard depending upon area temperature, and can be supplied in various colours. They sometimes sink during drying.

3. *Resin bound wood grain fillers*

These are the best of wood grain fillers as they are used under all surface coatings and proprietary branded products are produced and sold in auto shops and hardware stores. They are mixtures of lead free pigments and contain synthetic resins, amorphous silica and other extenders which are ideal for all types of surface coatings such as shellac, french polish, acid catalyst lacquers and cellulose. They are made in colours such as teak, oaks, mahogany, walnut, sapele and white (for liming). They are compatible with all surface coating finishes and will not react with the catalysts.

Most of these fillers come in thixotropic pastes which disperse into a creamy paste upon application to the substrate.

Catalysed wood grain fillers

The typical catalysed fillers, which are really wood fillers and wood grain fillers combined, and can also be used on metal. (They contain 13% styrene, hardener benzoyl peroxide). They are ideal where special surface coatings such as polyurethane and polyester, as well as pre-catalysed and acid-catalysed surface coatings are used.

It must be emphasized that when using any of the above mentioned products, care must be taken not to allow surplus filler to lay on the substrate as the hardened residue is almost impossible to remove, so wiping off the surplus and just leaving the filler in the grain is of paramount importance.

Removal of surface faults

In the preparation process great attention must be paid to scratches on the surface of the wood, because if these are not eradicated they show up as flaws on finished surfaces. Scratches can be removed in various ways – the easiest being the use of a cabinet scraper (if the substrate is solid wood) and finishing off with fine abrasive coated papers. If, however, the substrate is veneered then great care is needed.

Take, for example, a scratch on a veneered surface. There are two ways of eradicating this: first

Preparation of Wood Substrates

by filling with a wood filler and tinting, then polishing or applying a surface coating over. This may sound easy but in actual fact this could cause problems with colour and compatibility with the surface finish. The other way of dealing with the scratch is to apply warm water with a pencil brush to the scratch. Upon drying, the fibres will be raised and this can ultimately be fine sanded smooth. Sometimes the scratch can be very deep and the only answer to the problem is to replace the veneer or, in the case of a solid substrate, to fill the deep scratch with a catalysed wood filler, sand down flat and tint using compatible pigments, making sure that you use the solvent of the finishing coatings. For example, if using french polish, use pigments thinned with methylated spirits to mix with the pigment; if using a pre-catalysed lacquer then use pigments thinned with pre-catalysed thinners with a little lacquer to mix with the pigments. In other words the colours that you are applying, no matter how small the area, will react if not used with the compatible medium. The pigmented colours, however, must be of a very thin consistency or you end up with a convex area of pigment and medium.

Bruises can be steamed out using a small electric soldering hot iron pressed on to a wet cloth over the bruise area. This has the action of swelling the fibres of the wood, so bringing up the bruised area to the true level of the wood, which, when dry can be fine sanded with 240–320 grade garnet abrasive papers. Another way is to allow a small amount of warm water to lie in the concave area of the bruise and leave for a while. This has the same effect as the hot iron or a wet cloth, but takes longer to accomplish the desired effect of removing the bruise. Take care not to scorch the wood and surrounding surface.

Another problem on solid wood is 'knots' and those on pine are a great nuisance. No matter how a substrate is prepared and finished the knots will show up or protrude above the level of the surrounding surface area. If, however, the substrate is being painted or varnished these knots can bleed and ooze resin and so become tacky. This problem can be eased by treating the knot with a dewaxed shellac sanding sealer or stopping fluid. Apply two coats at least, using a small mop for the job, allow it to dry and fine sand before the surface coatings are applied. On pine furniture the knots will always show up sooner or later and it is interesting to note that even on new furniture finishes you can always see the knots breaking out of the surface finish, no matter how carefully the finishing has been undertaken.

When it comes to chipboard or MDF, new techniques must be employed. For example, if there is a large gash on one corner, or there is a small piece missing, the only remedy is to use a catalysed wood filler and build up the damaged area, flat down and tint in and also touch in to simulate the grain on to the filled area before applying the surface coatings. The beauty of these catalysed fillers is that they do not sink upon drying.

Health and safety check list

1. Have good ventilation in the work area.

2. Wear face masks if sanding and take care when handling certain products such as catalysed fillers not to inhale the odours or residual dust.

3. If bleaching has been carried out, great care should be taken when sanding.
4. If using sanding machines, make sure of adequate ventilation.
5. No smoking, eating or drinking in the work area.
6. Clean up the work area at the end of each day and properly dispose of any workshop waste.
7. Use barrier creams and wash hands thoroughly after using all chemicals.

CHAPTER THREE

Abrasives

Whether traditional or modern finishing techniques are used the most important product that a wood finisher has in his armoury is a range of abrasives, for without these vital papers and powders none of the many beautiful finishes can be achieved. This statement applies equally to the individual craftsman or craftswoman working by either hand finishing or the spray gun, or an industrial set-up using high technology with automatic spray systems and multi-pass drying systems. The result must be the same – a near-perfect, high-quality finish, and it is here that the correct use of abrasives is essential.

The early materials used were sand or ground shells glued to thin hides with animal glues, and later, ground brick dust, pumice powder and flint. Not only wood was polished but also brass, jewellery and stone. Up to the nineteenth century sand and brick dust were used by our cabinetmakers for sanding their products and these were glued on to sheets of paper, and crushed glass was similarly used. It is from here that we get the term 'sandpaper' which is still used today to describe abrasive papers, and also the methods used – hence – 'sanding', 'sander', etc.

Basically there are only two types of sanding papers, or, now more correctly, abrasive papers: cabinet papers which are used for smoothing wood, and finishing papers which are used on coated surfaces such as paint, lacquer, varnish, waxes, emulsions, etc.

Abrasive materials come as either natural or man-made and there are now many more of the latter which have replaced the more traditional glass and garnet from the wood finishing trade.

There are basically five types of abrasive papers:

1. Glass Paper

2. Garnet Paper

3. Aluminium Oxide Paper

4. Silicon Carbide Paper

5. Emery Paper

I mentioned flint as being abrasive, but due to health hazards it is illegal to use this material in Britain as the flint grains are thought to be a possible cause of silicosis.

Each of these abrasive materials has a specific purpose and use in wood finishing and newcomers to the trade are sometimes confused as to where and when to use these abrasives.

Glass Paper

This is one of the oldest kinds of abrasives and are mostly used throughout the D.I.Y. Trade. The glass is a mixture of brown and green glass from broken-up bottles which are crushed and graded, ranging from flour (very fine) to coarse. The traditional gradings are flour (which carries no grading number), followed by 0, 1, 1½, F2, M2, S2 and 2½ and there is also a flexible cabinet paper. The trouble is, nowadays, that due to abrasive papers being imported from all over the world, these grading numbers are sometimes confusing unless you keep to one manufacturer and become familiar with their products.

Garnet Paper

This is the most popular abrasive paper used by both the woodworker and finisher. It is normally a tawny red colour produced from a naturally occuring iron/aluminium silicon amalgam, and the best garnet rock is found in the mountains near New York. The beauty of this abrasive is that unlike glass the garnet breaks down during use and produces new cutting edges, so the paper lasts for a great deal of time whilst being used on the bench. The grading is usually from No. 80 (coarse) to 320 (very fine).

A word here about grading numbers: normally, the higher the number, the finer the grit. Thus the 240 grit is used for fine smoothing work whilst the lower grade of say 80 is a very rough paper. The problem, as previously stated, is that the firms manufacturing these products have their own grading standards and foreign manufacturers may differ. The cutting quality of the product also varies between different manufacturers and the user soon acquires a preference.

Aluminium Oxide Paper

These are some of the newest abrasives created by the fusion of rock bauxite by extreme heat. There are now various colours on the market: yellow, green, brown and white and the white is the near-purest form. The grading can be as low as 40 through to 150 which gives a fine finish on hardwoods.

It must be remembered that aluminium oxide abrasives produce a very sharp cutting paper and care must be taken when using it with a powered orbital sander on any hardwood as the cross-cutting action can show up as ugly scour marks, which can cause a great deal of wasted time in eliminating them in the preparation of a substrate.

These papers are used by both woodworkers and finishers in the preparation of most hardwoods that are resistant to the softer cutting edges of, say, glass and garnet papers. In my opinion, glass paper or even garnet, are almost useless on some hardwoods such as iroko and oak.

It must be stated here that all abrasive papers can cause damage if not used in a correct manner. The surface is, after all, being scratched by thousands of tiny points, the more the points the finer the ultimate so-called sanding effect or cutting action – hence 'cut', and it is paramount that the action of the sanding is used with the grain. Under no circumstances work across the grain as scaring will be the result, and these markings will be visible in any finish.

Aluminium oxide abrasive papers are ideal for all hardwood joinery and hardwood preparation and must be used with care. They should never be used on the softer woods and mouldings, or extreme damage can occur to the wood. When used with power sanding machines the clogging effect can be cleaned off with either a wire brush or one of the proprietory cleaning compounds now on the market.

Silicon Carbide Paper

These are the most sought-after abrasive papers for the wood finisher. They are not used by the woodworker for the simple reason that they are not intended for sanding wood as they simply clog up very quickly if used on bare wood. However, for the woodfinisher they are indispensible and 'cut' unlike many of the other mentioned abrasive papers. They are manufactured by the extreme high temperature fusing of silica, sand, coke and salt.

They are manufactured basically in two colour forms: the waterproof dark grey backing type which can be used with water as a lubricant, or, alternatively, with mineral or vegetable oil; and the dry white type which has a manufacturer's copyright name of 'Lubrisil' and is a product of English Abrasives Ltd. Various other manufacturers have similar products based on the principle of a dry silicon abrasive paper which is used without the addition of any further lubricant. The backing of this dry paper is not waterproof and will quickly break up if used with oil or water.

The waterproof silicon carbide, commonly known as 'wet and dry', is a dark grey colour, has a waterproof backing paper and is used extensively on de-nibbing lacquers, varnishes and paint on wood and metal substrates. If the correct grade of this paper is used it will also remove lacquer runs, cissing, overspray faults, insects, etc, from the surface coating film before pulling over or burnishing takes place. A little experience soon tunes you to the correct grade of paper to use. If in doubt use a fine grade first and work up or down the grits, say from 150 to 400 as a trial. The range of grits or grading goes from coarse (60) up to extremely fine (1200) depending upon the manufacturers coding.

Silicon carbide 'Lubrisil' is an excellent paper which is manufactured with a chalky, talc-like lubricated surface of a light greyish white colour. It is used dry and is ideal where water must not be used on a substrate. The grading of this paper is normally 80 (coarse) to 400 (very fine). It is also ideal for cleaning up old brass fittings such as hinges and locks and for metal body work where the substrate has to be kept dry for some reason. This product is excellent for finishing wood on the lathe whilst the turned item is at speed. There are some occasions where a dry abrasive paper is better than the water or oil wet and dry lubricated type. In traditional finishes, such as in the french polishing technique, it is better than 'wet and dry' papers.

Emery Abrasive Paper

These papers are mostly used on metal finishing such as brass, iron, steel etc, and they are supplied in two basic forms: emery cloth backing and emery paper backing. The raw emery product comes from Turkey and Greece and it has a high corundum content. The cloth backing is of a cotton/denim-type material or a combination of cotton and man-made fibres such as polyester that gives it strength, whilst the cheaper types of emery abrasive sheet are backed by strong paper.

Tungsten Carbide Metal Sheets

These are synthetic minerals affixed to thin flexible metal sheets and can be used on wood, metal, plaster, fibreglass etc. The great advantage of these is that they last some considerable time before wearing out. They are ideal for auto body profile work, i.e. sanding catalyst-type fillers.

Steel Wool

This very important material is indispensible to the wood finisher and is used in traditional and modern wood finishing techniques.

In traditional polishing very fine steel wool can be used to flat down a surface such as in french polishing and can also be used to produce a fine semi-gloss finish. In varnishing, flatting down with steel wool will produce a fine satin finish, and in antique restoration steel wool is often used with wax polish to give a fine mature 'patina' to fresh french polish. To clean old, worn shellac surfaces fine steel wool used with a little reviver removes all dirt and grime. It can also, under certain conditions, be used to remove ugly white rings from french polished surfaces by using the finest of steel wool and a little wax polish, and on modern surface coating films fine steel wool is used to produce that fine quality finish over a pre-catalysed lacquer when a semi-gloss effect is required.

A summary of uses for steel wool
0000 and 000 gradings are used for all fine work and will not show scratches.
00 and 0 are for general household use – rubbing down paintwork, varnish, etc.
No. 1 and No. 2 are ideal for cleaning machinery, tools, etc.
No. 3 is used for removing rust from any metal substrate such as metal down-spouts, car body components and garden tools.
Nos. 4 and 5 are the coarsest and are ideal for using when stripping woodwork in conjunction with a chemical stripper or for rubbing down parquet floors, etc.

Most manufacturers supply steel wool by the 1 lb or kilo rolls.

Abrasives

Pumice Powders

These are produced from natural lava deposits ground down to various grits, are off-white in colour and supplied in various grades. To the wood finisher they are ideal for producing the piano finish technique when used with a beezer in french polishing, and are also used in revivers for cleaning very dirty polished surfaces, or to flat down paint or varnish surfaces between coatings. Another use of pumice is in distressing techniques to wood and to finishes.

Rottenstone Powder

This product is sometimes known as Tripoli Powder and is much finer than pumice. It is siliceous limestone powder, pinkish in colour and is used in most burnishing or buffing pastes such as cellulose and pre-catalysed burnishing pastes, burnishing creams and car polishes. It can be used mixed with a little wax polish, linseed oil, or mineral oil to burnish varnish or a french polished surface.

French Chalk

This is a mild acid magnesium metasilicate – a fine talc-like white powder which is used in mild abrasive creams, revivers and haze removers. It is ideal for drawer slides, table slides, piano actions where wood has to rub against wood, or as a lubricant when sprinkled upon dance floors.

Vienna Chalk

This is a fine precipitated soft chalk, off-white in colour and is used for the final removal of oil deposits from a surface when used in conjunction with mild sulphuric acid by the french polisher in the technique known as the 'piano acid finish'.

Backing Materials

Regardless of whether the abrasive grit is glass, garnet silicon carbide, aluminium oxide or emery, it requires a backing paper or cloth sheet and therefore the bonding of the adhesive and the 'grit' is not as simple as it sounds. For normal glass papers, animal glues are used and here you can smell the typical 'boiled glue' aroma of the product, but when it comes to the other sophisticated abrasives such as silicon carbides etc, various combinations of adhesives are used such as cold-cure set resins over animal glues, or various layers of resin cold-cure adhesives provide a waterproof backing. A good plan when using these waterproof papers is to keep most of those being used in actual water; this has the advantage of cleaning the papers until they are next used, and another point here is that it softens the papers which makes for easier use. I make it a workshop plan to keep all slightly used and worn abrasive papers in this way – keeping them in the categories: garnet, aluminium oxide, silicon carbide etc, all in separate containers. It is a rule of mine that only abrasive papers which are completely worn away are discarded. It must be remembered that abrasive materials are not cheap any more, so cost is the drive for thrift. It is also a wise plan to keep the used sanding belts or orbital sander pads and cut them into small pieces and re-use them for hand sanding.

Most people rarely look on the back of a piece of abrasive paper sheet. However, paper backings are marked with numbers and coded by weight. The letters A, C, D and E refer to weights of the paper which are measured in grammes per square metre. Also the actual grit and name of the product is also marked, such as 'Garnet Paper' A.OP 7/0–240 – this means that it is a light paper and the grit is 240 (very fine). Due to imports these numbers do vary and the important point to remember is the grit number – you can feel the weight of the paper yourself.

It is a good guideline to note that for all hand sanding, a lightweight paper is an advantage, but for machine usage a heavyweight paper is the order of the day.

In the main-line stores and the high street D.I.Y. shops, very little change has taken place in the public knowledge of abrasive materials and one can find an abundance of glass papers and wet and dry silicon carbide papers and generally nothing more. I think in this day and age it is a lamentable neglect on the part of these stores not to stock a wider selection from the full range of abrasive materials now available, and this applies to most countries throughout Europe (including Britain) and N. America.

It is interesting that in the manufacture of coated abrasive papers, glass paper is still made by passing powdered glass (by gravity feed) onto its coating of adhesive, whilst all the other coatings are produced by electrostatically forcing the grits onto the previously bonded papers. Using the latter method the grits or grains are fixed sharper onto the bonding adhesive and have greater cutting power than say glass paper, which is an advantage regardless of the particular abrasive material being used.

A summary of Abrasive Papers

Glass Abrasive Paper
Ideal for the D.I.Y. trade and non-commercial wood work.
Grades: Flour, 00, 0, 1, 1½, F2, M2 and S2.

Garnet Abrasive Paper
Ideal for general joinery and cabinet making.
Grades: 80, 100, 120, 150, 180, 220, 240, 320.

Aluminium Oxide Paper
Ideal for woodwork dealing with hardwoods such as iroko, oak, rosewood, etc.
Grades: 40, 50, 60, 80, 100, 120, 150.
Abrasive Sanding Belts: these are generally in widths of 100, 75 or 63 mm, and in lengths of 610,

560, 480, 404 mm. (Approximate imperial measures 4″, 3″ or 2½″ and lengths 24″, 22″, 19″, 16″). These sizes may vary according to country of origin.

There is a sanding belt cleaner for this type of abrasive available.

Silicon Carbide Abrasive Paper (Wet and Dry)

Ideal for all modern finishing coatings such as cellulose, pre-catalysed, acid-catalysed, polyester, water-bourne lacquers, varnish, paint, surface coatings on metal body work, french polish and shellac film surfaces and for removing spray faults. It is not suitable for bare wood.
Grades: 60, 80, 100, 120, 150, 180, 220, 240, 600, 1200.

Lubricated Silicon Carbide Abrasive Paper

Ideal for all modern wood surface films including french polish and shellac.
Grades: 80, 100, 120, 150, 180, 220, 240, 280, 320, 400.

Bench Sander

In the commercial world, sanding is carried out using the latest developments in modern technology. One such development is the 'Sanding Bench'. This is a metal bench with a hard green felt fixed to the top of the bench where two or more operatives can work simultaneously. Under the bench is incorporated a centrifugal fan which draws air down from the top of the bench working area. This powerful down-draught air stream draws all the sanded wood dust mixtures down through the felted bench top and it is collected into bags and thus the operatives work on hand sanding in a dust-free atmosphere. The system also keeps the dust extraction and noise down to a minimum and conforms to UK COSHH regulations.

Shaft Flexible Sander

This is a new-comer to the wealth of inventions and has a hand-held shaft to which a rotating head made up of cloth strips is attached. The system is constructed for sanding substrates of awkward shapes such as moulded door panels and for de-nibbing the most difficult profiles. They can also be used for removing old coatings, removing rust, and for general cleaning and are supplied in various grits from 600 for fine work to 60 for the coarsest work.

Fig 1   Hand held tool for sanding and finishing
This is a pneumatic lightweight one-hand tool used for sanding larger items where it is an advantage to have one hand free. It comes with a protection shield which is easy to mount/dismount. The trigger action gives the possibility for a slow start.
Photograph courtesy of Gibbs Finishing Systems

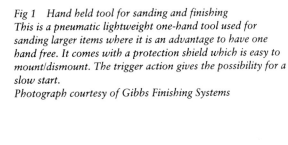

Fig 2   Brush-backed sanding head
For contour sanding on hard and soft wood, veneer and MDF. The very flexible abrasive is pressed gently against the surface to sand it. This is a modular system based on a 50mm (2") hub.
Photograph courtesy of Gibbs Finishing Systems

Fig 3 Oscillating brush wheel
Made for brush sanding and finishing. The brushes can be made for special purposes or to specifications. This is a modular system based on a 15mm (5/8") hub. A bundle of brushes is mounted on the hub and can be built together to almost any length. The hub can be refilled when worn out.
Photograph courtesy of Gibbs Finishing Systems

Fig 4 Automatic Universal Sander
Unique dual head system combined with the 'Quick Disc' which significantly increases the efficiency of the finishing operation. This system will secure a 100% symmetrical sanding.
Photograph courtesy of Gibbs Finishing Systems

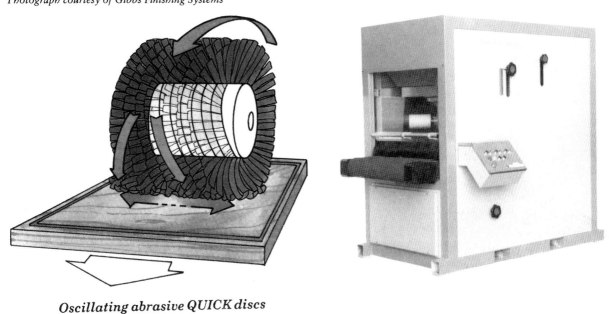

Oscillating abrasive QUICK discs

Disc Sander and Orbital Sanders

On the hand-held machines the disc sander and orbital sander can now be combined in one tool. The ingenious combination of rotating and orbital action makes these power tools produce very fast results in sanding. They have, of course, built-in dust extraction and variable speed control.

Universal Sanders

These sanders are ideal for the small and medium sized workshops. They will perform edge and face sanding operations and thus finish rebates and profiles. Firms producing kitchen fixtures, desks, general furniture, etc, find this type of sander indispensible.

Wide Panel Sander

These machines can be operated by a single skilled operative and the substrates, which can be wide panels for doors, wardrobes, etc, are simply guided through the sander using the applied power. No hand pressure held pad is used with this system.

Programmed-Controlled Sander for the Computer-Controlled Factory

These top-of-the-range sanders are for mass production sanding and can be adjusted for such things as tolerance compensation, belt-cutting speeds, sanding elements and general adjustments which are all carried out automatically.

The dust-extraction outlet is close to the sander and these machines have precision sanding actions suitable for veneers, lacquered films, and a rigid pad for the sanding of small items such as rails of solid wood.

These are the most perfect sanding machines that modern technology has yet produced – a far cry from the cork block hand sander of yesterday!

Below is a small selection of other industrial machine sanders, many of which are very expensive, showing the scope of machines that smooth wood and process-finished surfaces today.

Twin contact pad highspeed sanding machines.
Open-ended belt pad sanding machines.
Open-ended over and under belt sanding machines.
Open-ended over and under belt pad sanding machines.
Single contact speed sander with dust extractor sanding machines.
Single contact speed sander.
Twin contact pad, high speed open-sided sanding machine, high speed triple.
Drum sanding machine with dusting roll.
Single contact underside speed sander.

Abrasives

Fig 5   De-nibber (DN600)
Designed for smaller manufacturers to de-nib panels up to 600mm (24") wide, it is fitted with twin counter-rotating spindles set at 45° to the feed path. Flexible sanding blades reach into grooves and corners for such items as fielded door panels. The feed bed is adjustable for height.
Photograph courtesy of Gibbs Finishing Systems

Health and Safety in dealing with abrasive materials

1. Always use a sanding machine with a dust extractor attached. (It is not the actual abrasive material here but the substrate which is being sanded that is the hazard – e.g. teak and rosewood dust).

2. Never use steel wool near a grinder (one small spark can set alight steel wool and thus become a fire hazard). N.B. Steel wool either glows and gets red hot or flares.

3. Never use steel wool on a turned item on a lathe whilst at speed, as it can wrap round a finger quite easily.

4. Always keep power sanding machines as clean as possible.

5. Never tear steel wool – cut the amount required with a pair of scissors.

6. Use gloves if possible when handling steel wool.

7. Have good ventilation when sanding with power sanders, or better still, have extraction equipment.

UK industrial users must conform to the COSHH regulations (The Control of Substances Hazardous to Health); US industrial users must conform to the OSHA regulations (Occupational Safety & Health Administration).

N.B. Wood dust can cause respiratory illnesses because the dust could contain various types of allergens and toxins. Individuals respond differently to these hazards so care is required in all matters of sanding.

CHAPTER FOUR

Wood Infestation

Sooner or later anyone concerned with wood finishing, antique restoration or general joinery will come across the scourge of wood – the wood-eating insect! If, however, your substrate is infected with these little horrors then you cannot proceed until the menace has been eradicated.

The first tell-tale signs of infestation in a piece of furniture is the fine creamy-white powder which is found either by the legs of furniture or under carpets – the worms actually eat the paper that has been laid under the carpets as they come out of the floorboards. This powder is called 'frass' and is actually the excrement of the woodworm grubs. This discovery is all too common in property and in furniture today. If an attack is dealt with promptly then it is likely that very little harm will have been done; but if nothing is done, then very quickly your furniture and floorboards will be infested with hundreds of wood-eating grubs.

Wood-eating insects fall into two groups: firstly, insects that inhabit the joinery and carpentry in buildings, such as beams, rafters, door frames, doors, floorboards, joists etc, and in this group the most commonly known is the death-watch beetle (*Xestobium rufovillosum*). Signs of infestation by this insect is easily identified because the holes are much larger than a common woodworm hole, and attacks are common in many old buildings and church towers and they are particularly active in areas where old oak is prevalent.

Another common beetle that inhabits joinery is the Power-post beetle (*Lyctus spp*), and also the House Longhorn beetle (*Hylotrupes bajulus*). These beetles cause great damage, taking the 'guts' out of the wood which leads to its ultimate collapse.

The second group are those which attack furniture and of these the most widely known is the Common Furniture beetle (*Anobium punctatum*), and this is what is normally found in dining-room, bedroom and other household furniture.

Between May and August the adult female beetle flies around looking for a nice dry undisturbed corner to raise her family. She then lays an average of about forty eggs which hatch into larva in about four to five weeks, and these will immediately bore into the wood where they continue to tunnel for between two or three years, feeding on your precious furniture. At the end of this stage they will bore to just below the surface of the wood, pass into the pupal stage and finally emerge after about six to eight weeks through flight holes – another characteristic sign of infestation. As full-grown insects they begin their sinister work all over again.

In my many years as a wood finisher and furniture restorer I have seen infestation by the common furniture beetle in such woods as oak, pine, mahogany, walnut, plywoods, and even in such woods as rosewood, bamboo, boxwood and pitch-pine. However, some woods have a built-in chemical immunity to these pests – woods such as teak, cedar and coconut wood, American redwood, iroko, eucalyptus, ebony to name a few, but it is a foolhardy person who states that woodworm never attacks such-and-such a wood! Wood-boring insects have no respect for timber. Furniture like any man-made object requires attention, and in addition to cleaning and polishing should be inspected annually for any tell-tale signs of insect infestation and any such infestation should be treated as soon as possible.

Treating furniture which has infestation by woodworm

Remove the furniture to, say, a garage, and place a polythene sheet under it. Remove drawers, shelves and any loose parts of the piece. With a vacuum suction tube and a stiff 50mm/2″ paint brush remove all traces of dust and any visible insects.

Using a proprietary anti-woodworm fluid, paint all over the bare or non-polished sections of the piece such as under shelves and under the base and top sections. It is always a good idea if at all possible to turn the furniture upside down first as it is far easier to treat this way. If there are visible signs of woodworm flight holes, then you must inject or drop the anti-woodworm fluid into these flight holes. I use a very thin hair pencil brush for this tedious job, but it is the only way to be sure – the fluid flows down the flight cavities and reaches the worms. The fluid combined with the odour kills off all live infestation.

If the infestation is very severe – say in a table leg – there are two ways of treating this. Above the line of infestation holes, drill out with a power drill, 6mm/¼″ holes to a depth of say 50–75mm (2–3″), then, using a fine funnel, pour in the anti-woodworm fluid until the wood will take no more. It is best to stand the leg within a tin to collect any waste fluid. This way and this way only will you be sure that treatment is carried out – the object of this system is to kill the worms and sterilize the wood. When dry, the holes you have made can easily be filled by plugging with dowels using a matching wood.

As an alternative to the above method, legs that are slightly infested can be stood in a tin of fluid which will also be drawn up into the fibres of the wood. Of course, if the severity of the attack has caused the part to become structurally weak, then this means that a section of the wood may have to be cut away and a new piece spliced in.

After carrying out this treatment, cover the whole piece with a plastic sheet to form a bag which acts as a fumigation tent and leave for at least twenty-four hours.

If this job is carried out during the summer months the piece can be placed outside in the air to dry off after fumigation has taken place. When the furniture has been returned to its normal position in the room, and if it has a shellac finish, wax all over using a polish with anti-woodworm ingredients in it. If the furniture has a lacquered finish, wipe it over using a cloth wrung out in a mixture of warm water and a little clear white vinegar (acetic acid) and when dry polish with a dry cloth. If it is a full gloss lacquered finish you can finish off by using one of the non-wax aerosol plastic cleaners for this job. In this way all deposits of woodworm fluid are removed.

Wood Infestation

Fig 6 Common Furniture Beetle
Anobium punctatum

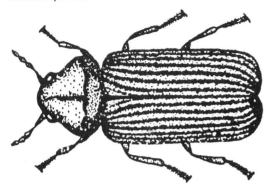

Fig 7 Infestation of the common furniture beetle in an antique mahogany table

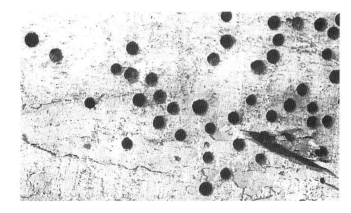

Fig 8 The large flight holes of the Death-watch Beetle in infested structural oak

Photographs courtesy of Rentokil Ltd.

Treating joinery which has infestation

When woodworm infestation is found in such items as floorboards or in roof areas, this is a different kettle of fish from woodworm found in furniture. Here one cannot feed each individual flight hole with an anti-woodworm fluid and therefore another technique has to be employed. In floorboard infestation the procedure is to lift, say, one in four floorboards and inspect for any serious damage. If only visible signs of flight holes occur, then a good clean-up is necessary to remove any rubbish such as shavings etc, and an application of an insecticidal fluid to all exposed areas will suffice. Here you will require a spray attachment to spray under as far as you can beneath the floorboards you have not lifted. If by any chance new boards have to be fitted make sure that they are also treated, even though the wood is new. When the job is finished spray or paint the fluid onto the whole of the floor surface and lay a sheet of polythene over the whole area. This has two purposes: firstly it prevents the odour from rising to the other parts of the house, and secondly forces the fumes down under the floor area where it is needed and can do most good.

When flight holes are visible in painted woodwork this requires stripping off to expose the bare wood and then the flight holes can be injected and the wood sprayed or painted with an anti-woodworm insecticidal fluid, and when dry, repainted.

In roof areas it is a good plan to first clean the area using a powerful industrial vacuum cleaner and brush to remove all dust, cobwebs etc. The whole area should then be sprayed using a strong insecticidal anti-woodworm fluid. It must be mentioned here that firms specializing in this kind of work use very strong fluids which are not normally obtainable by the general public, but there are many brands of anti-woodworm fluids on the market that will give good service for the self-help D.I.Y. enthusiast. When it comes to houses, it is always better to have specialist firms to do this kind of very dirty work as they normally give a guarantee for thirty or more years, which could also be a selling advantage to the owner.

If work of this kind has to be carried out in your home, then it will be advantageous to you to have the work done whilst you are away or out of the house for a few days to allow the fumes to disperse (particularly if any of the family suffer from asthma or other chest complaints) although these days, some of the woodworm insecticides have very little odour.

Cosmetic filling of flight holes

In furniture, once the treatment has been carried out you then have to deal with the unsightly flight holes made by the worms. If traditional finishes are used then coloured waxes can be used to fill these cavities before polishing over. If bare wood has been infected then modern glues mixed with fine sawdust can be used, or, alternatively, any of the pre-mixed wood fillers can be used, but these tend to be rather soft and sink when drying.

In the case of modern finishes such as full gloss or semi-gloss finishes such as acid-catalysed or pre-catalysed lacquers, cellulose, or even varnish, a very hard filler must be used. On no account use any waxes under these finishes. Catalysed wood fillers (the kind used on car body repairs) are ideal as they do not shrink on drying and they can be sanded, coloured and lacquered over without any

Wood Infestation

Fig 9 Infestation by both Common Furniture Beetle and Death-watch Beetle causing the complete breakdown of the wood fibres in this piece of oak.

problem to the finish. Care must be taken in touching up the filled holes after sanding and also make sure that the pigment is mixed with the compatible finish – e.g. if using cellulose then use this lacquer to mix with the pigments to obtain the perfect match, when the filling and polishing or lacquering of these flight holes will not be visible after the job has been completed.

Fluids obtainable to treat for infestation in furniture and joinery

Common woodworm beetle (*Anobium punctatum*) – fluid to use is an anti-woodworm fluid (one coat).
House longhorn beetle – fluid to use is an anti-woodworm fluid (two coats).
Powder post beetle (*Lyctus* species) – fluid to use is an anti-woodworm fluid (one coat).
Wood boring weevils (found in damp or decaying wood) – fluid to use is a dry rot fluid, but you must also take steps to remove decaying wood and replace with new.
Death-watch beetle (*Xestobium rufovillosum*) (Holes are much larger than those of the common furniture beetle) – fluid to use is an anti-woodworm fluid, and you must also take steps to eliminate any excess moisture.
   In addition to insects eating away wood, a number of fungi feed on the naturally occurring cellulose in timber and can reduce it to a brittle and useless condition. The dry rot fungus (*Serpula lacrymans*) is the most serious of wood destroying fungi which can be treated by using a proprietary fluid specifically formulated for the treatment of fungi.
   Note: Wet rot is not as serious as dry rot, however it requires careful treatment at the earliest

Fig 10 A typical example of dry rot (Serpula lacrymans) which shows the complete break-down of the wood fibres rendering it useless.

stage to avoid subsequent costly damage. The growth of wet rot will only take place in buildings where infested timber has a moisture content greater than 50/60%.

Furniture can be treated for woodworm infestation either by using one of the proprietary brands or by making your own.

Formula for a general mild cleaning fluid for treating furniture suspected of infestation

a) To five litres of white spirit (mineral spirit) add 100 ml of strong acetic acid (80%) and 100 ml of mineral oil. Shake up and apply to the bare wood. Paraffin oil (kerosene) can be used instead of white spirit;
   or
b) To five litres of white spirit or paraffin oil add 100 ml of light creosote oil (this really disinfects any piece of furniture but the slight snag here is that the odour can linger for a while;
   or
c) To 2½ litres of warm water add 100 ml of mild disinfectant, and a drop of detergent. This is only to be used on bare wood and must not be applied to polished surfaces. Apply during a very warm day so that evaporation is swift.

As prevention is always better than cure it is a good idea to sterilize furniture annually or when it has been in store for a few weeks or months simply by applying anti-woodworm fluid by brush or spray

Wood Infestation

onto the bare parts of your furniture, and this should help to ensure that no infestation takes place.

Health and safety check list when using any woodworm or dry rot fluid insecticides

1. Always make sure there is plenty of fresh air (i.e. adequate ventilation).
2. Wear a face mask to cover mouth, eyes and nose when spraying or using fluids in a confined space such as a roof or floor void.
3. Wear plastic gloves when handling fluids.
4. Remove pets, fish tanks etc whilst the area is being treated.
5. Keep children away from the areas being treated.
6. Make sure that people who suffer from chest complaints are not around when the fluids are being used.
7. Make sure that old clothing or, more correctly, proper working clothes are worn.
8. Make sure no naked lights – gas heaters etc are on at the time of application.
9. Make sure that no one is cooking, eating or drinking in or near the area during treatment.
10. Wash all working clothing separately after treatment has finished.
11. All electrical connectors, switches etc should be covered with masking tape whilst the fluids are being applied. In the loft any water tanks of the open type must be covered up with polythene (polyethylene) sheeting until the treatment is completed.
13. Make sure that the instructions on the manufacturer's containers are followed.

CHAPTER FIVE

Bleaching

In general joinery work or furniture making there are times when varying woods are used and although the supply may be from the same source and same batch, when made up as completed work, areas of the wood, as far as grain and colour are concerned, can look completely different. If you simply use a unifying stain, whether it be an oil, spirit, water or mixed solvent stain, the unbalanced colour can look cheap and nasty. This can be disastrous in quality finishing, and it is here that 'bleaches' come to the rescue of the wood finisher.

Aesthetically, there are purists who object to any form of bleaching the natural colour out of wood simply to lighten it, and you could also say that staining wood and altering the natural colour is also within this net, or at least debatable. Every woodworker must decide for themselves which working practises they find acceptable.

A great deal of nonsense has been written over the past years by people who are not quite sure of their ground, and, as a result, false information has been reproduced by writers and also by some manufacturers of bleaches on their instruction labels, which have been amended by some only in the past few years.

Although it is greatly used by antique restorers, kitchen fitting firms and contract wood finishers, it must be emphasized that bleaching can only be carried out to the natural substrate and therefore any existing surface coating must be removed first by normal stripping techniques and allowed to dry out completely.

Why do you need to bleach wood? The answers are various: to change the natural tone to a lighter one; to 'age' new veneers prior to fitting for antique furniture repairs; to remove ink and rusk marks and ingrained dirt which cannot be removed by stripping; to distress colour, particularly in reproduction furniture – to name just a few.

The bleaching of a wooden substrate takes place by a chemical process of oxidation or the reaction of hydrogen either with the chemical within that wood or alien chemicals applied upon it.

Now, at this stage I must digress and point out that there is a subtle difference between the terms 'bleach' and 'fade'. Any furniture retailer knows that during the summer months he must not affix labels to or place any items on the furniture within the window area which catches the full rays of the sun as this can lead to a very serious pronounced patch of fading around the contour of the item placed upon the furniture, and this is a most difficult job for a polisher to eradicate. Similarly, in many households certain furniture, such as teak, quickly fades in strong sunlight and great attention to the positioning of furniture within a room and to the drawing of curtains to minimize direct sun rays is vital for the retention of the colour of the furniture. This applies not only to wood but to fabrics as well. However, not all woods fade: as any kitchen fitter will tell you, the common pine so popular today actually darkens with sunlight. To test this statement, simply place a picture on a new pine-panelled wall and remove it a few months later: the area around the picture will be considerably darker than the area behind the picture. Other woods that actually darken are oak, iroko, yew, beech, ash and sycamore, and woods that fade are birch, teak, white oak and African mahogany. Western red cedar is one that first darkens and then fades over the passage of time. Notice modern timber bungalows built by specialist firms in this field and see how, after a few months, cedar shingles darken when newly fitted and then turn a lovely silver grey due to the effect of sun and rain. To the wood finisher therefore, fading is the action of the ultra-violet range of the sun's rays reacting on the wood and the finish, whilst bleaching is a chemically produced finish brought about by the reaction of alkalis and acids upon the natural chemical compounds contained within the cells or structure of the woods in question. Both really are comparable, however, in the resultant finish effect.

To the wood finisher who uses bleaching products these practical problems are known and understood, but are often misunderstood by the amateur. The practical application of a 'bleaching' technique is not as simple as it may appear and, for the novice, many latent pitfalls can be encountered and serious problems can occur later, well after finishing has been completed.

The various bleaches available to the wood finisher on the market today are few, which is a great advantage. The weak household bleach ideal for treating the drains or cleaning the enamel sink has little use for the wood finisher for whom the original and more traditional bleach is oxalic acid, which is supplied in crystal form and diluted with water. Remember, ALWAYS ADD THE ACID TO THE WATER – not the other way round. A stock solution can be made up of 1 lb of the crystals in 1 gallon/5 litres of water which must be fairly warm or the crystals will not dissolve. This standard mix produces a mild bleach which is ideal for removing burns, ink, rust and water stains on most woods. When the bleaching action is complete use water by drenching to neutralize or, alternately, some traditional polishers use a weak solution of borax or soda to aid neutralization. If the latter is used it must also be washed later with water as there must be no deposits of the bleach left on the wood as this can cause problems later to the substrate finish, and can show up even months after finishing has taken place.

The most popular method of chemical bleaching today is the two-pack system. This consists of two plastic bottles: the first which is called, literally, Solution 'A' or 'No. 1', is a strong alkaline such as caustic soda or ammonia, which acts as a degreaser on the wood. This is applied upon the surface of the wood and left for approximately 10–20 minutes. The second bottle known as Solution 'B' or 'No. 2' is hydrogen peroxide (acid) and is now applied to the still-wet substrate and a foaming action takes place as the two chemicals react with one another, and the bleaching action takes place as the alkaline and acid battle it out on the substrate. This reaction could take between a half and one hour depending upon the wood, and it is difficult at first to judge the timing but this will come with a little experience. When the time has come to end the bleaching process the substrate must be washed down with clean running water using a hose pipe and scrubbing brush and be allowed to dry out.

In the past, a great deal of nonsense has been written about using vinegar or mild acetic acid to neutralize the reaction but all this does is to use a mild acid on top of the stronger acid and is therefore a waste of time. To prove my point, most manufacturers have now removed the 'vinegar' neutralizer method from their bleach bottle labels. Instead, as I have just said, use plenty of water with the help of a pressure hose supply, and a scrubbing brush to clean the substrate surface of unwanted chemicals, then allow it to stand for about twenty minutes wet and *re-wash again*. The surface should then be dried out as quickly as possible with the aid of absorbant paper towels and left for at least twenty-four hours either in a well-ventilated room or, during the summer, in hot sun. If you try and refinish when moisture from the bleaching process is still within the substrate, then no end of troubles can occur, particularly if using lacquer or shellac finishes. Great care must be taken in washing out all traces of bleaching chemicals from the substrate with water – this I cannot emphasize enough!

Super Bleach

This is now a popular bleach which is a concentrated hydrogen peroxide solution (27.5% w/w), which has to be mixed with .880 ammonia which is a very concentrated and dangerous material to handle, thus great care is required when handling this product as it can cause drastic respiratory problems.

The two solutions must be mixed in either a glass, enamel, or poly container in a mixing ratio of approximately one part ammonia to 20 parts bleach. Upon mixing, great care is required as the two solutions mixed will effervesce, so room in the container must be allowed for this reaction. The

Fig 11   The result of bleaching oak

effect on wood is quite staggering and normal washing with water is carried out after bleaching to neutralize the effect.

Extra health and safety protection procedures are essential when handling this product.

The concept of bleaching is complicated by the fact that not all woods, be they softwoods or hardwoods, respond to even the strongest of bleaches. These are frequently the close-grained woods such as Burmese teak, cedar of Lebanon, Western red cedar, ebony, rosewood, satinwood and eucalyptus. Woods that are easily bleached are the open-grained species which respond very easily to chemical bleaching such as ash, elm, sycamore, maple or chestnut. In general, open-grained woods react to bleaching and close-grained woods do not, but woods such as mahogany and beech for example, which are both close-grained, do respond!

In modern finishing methods today light woods are the vogue. Woods such as ash, pine, poplar, beech, bird's-eye maple, elm, cherry, lime, sycamore, satinwood or zebrawood are very light coloured in their natural state and applying finishes sometimes darkens them. However, due to the advance in chemical solvents and non-solvent lacquers, finishes can be applied which will have no effect of darkening the woods and the new breed of water-borne lacquers are ideal for these types of wood. However, even some of these woods can be made lighter by resorting to bleaching.

Bleaching is also used to remove deep stains from wood such as ink or rust marks from steel screws on oak. After the surface coating has been removed by chemical stripping, the area so marked can be treated with oxalic acid and left wet for between one and two hours until all traces of stain in the wood have disappeared. This will vary according to the depth of stain. The treated area should then be washed with water and allowed to dry out, and then sanded down with 240 grade

Before   *After*

Fig 12   The result of bleaching elm

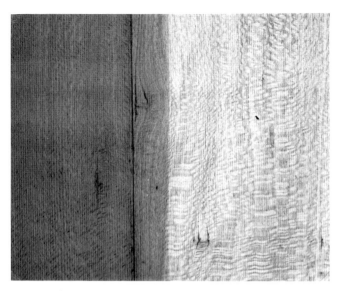

Before   *After*
Fig 13   The result of bleaching beech

garnet paper after which the substrate will be ready for any refinishing.

Pine, after being stripped using caustic soda, darkens in colour and to lighten the wood the two-pack bleaching method is applied.

Yet another use of bleaching is to clean up a substrate before refinishing. Take the case of a set of Windsor chairs, all of different woods – after stripping had taken place they were found to be of varying shades of beech and elm, and dirt and grime were still engrained in the wood. By using the two-pack bleaching method the woods were made uniform and ready to take a compatible colour and finish.

It must be remembered that any chemical bleaching deposit left on the surface of the substrate which may not even be visible can react with some surface finishes, particularly the cellulose-based type and acid-hardened lacquers, either upon application or in the curing, so great care is required.

In antique restoration work, bleaching is used on veneers before re-fitting to simulate age and fading so that when re-polishing is carried out the new veneers blend in to the original section of the piece.

It must be made clear that no bleaching should be done on veneers originally fixed with organic glue which requires heating in a glue kettle. Other veneers fixed with modern glues such as urea formaldehyde resin or one of the new water resisting polmerised vinyl acetate adhesives (PVA) *can* be bleached. It is always prudent to carry out a test area first before proceeding with the whole bleaching process to make sure that the veneer glue is non-reversible.

To sum up, the bleaching process is used for the following reasons:

1. To change the natural dark tone of wood and to make lighter.

2. To remove stains and ingrained dirt and grime.

Bleaching

Application of caustic soda
(The reaction of the two chemical compounds)

Application of hydrogen peroxide

The finished result
(An evenly balanced colour of wood ready for re-finishing)

Figs 14–16 The three stages of bleaching wood using the two pack system

3. To simulate age when refitting new veneers and other woods to antique furniture.

4. To unify the colour of varying woods prior to staining.

5. To lighten pine after being stripped with caustic soda.

6. To distress colour, particularly in antique furniture restoration work.

Health and Safety Advice

Before anyone considers using chemicals for bleaching a few basic health and safety precautions must be followed or serious burns to the skin can occur. These are as follows:

1. Wear old clothes with protective plastic aprons (or rubber), a good pair of strong gauntlet rubber or plastic gloves, goggles and gum boots. Use a face mask to cover the eyes, nose and mouth when using .880 ammonia.

2. Carry out the whole process out of doors near a grid with good drainage but not near grass or other vegetation.

3. Apply the chemicals by using a 'grass' brush which is made of tampico and other vegetable fibres which are resistant to strong chemicals.

4. Use an old enamel pan or glass jug to hold the chemicals whilst applying them.

5. Have a hose-pipe with pressure attachment connected to a good water supply as all traces of bleach must be washed out of the wood. It also helps to scrub the substrate using a hard scrubbing brush (not nylon) whilst washing down.

6. Store all chemicals in a cool dark place – never in sunlight and never return semi-used chemicals to their original containers.

7. Ensure that children and/or pets are well away from where you are carrying out the bleaching process.

8. Have a bucket of clean water and sponge handy for use if any accidental spillage goes onto your skin.

9. If solid pieces of wood are to be bleached, such as a table-top, make sure that a wooden cleat is screwed underneath across the grain. This will prevent bowing whilst the wood is wet and can then be removed after the item has dried out.

10. Never dry-sand a substrate which has been bleached and allowed to dry before neutralizing with water as the dust can be dangerous. The substrate must be washed down in the normal way – dried – then sanded.

It must be emphasized that bleaching chemicals are toxic and dangerous and therefore must be treated with extreme care, and providing these guidelines are followed the process is quite safe.

Bleaching 45

Fig 17 Part of the standard equipment required for bleaching wood. (Note: Protective clothing, boots and goggles are also required.)

CHAPTER SIX

Stripping and Cleaning

Wood finishers, be they contract polishers, furniture restorers, french polishers, construction finishers, bedroom and kitchen finishers, D.I.Y. enthusiasts etc, all use some technique of removing a hard surface coating film off a wooden substrate. Antique dealers and some television presenters shake in horror if the word 'strip' is used where antique furniture is involved, but it all depends upon the surface. Nobody in their right mind would strip an antique or part of an antique just for the fun of it, but furniture, like any man-made object, will not last forever and may require at a certain time in its life what in the car trade is called 'servicing'. Wood and its finishes, no matter how well in the first instance they were constructed, will not outlive time. On the other hand, furniture and joinery become damaged by either fire, water, bad usage, children and the rest of what mankind can think up, and therefore on furniture surfaces and internal doors, windows and staircases these hazards, plus the effects of the passage of time, damp climates or excessive central heating, all prove too much for the thin surface coatings on our furniture and woodwork.

So the time comes when all or part of a section of woodwork construction or a piece of furniture has to be stripped of its surface film, so that re-finishing can take place to restore the wood to its original finish. I can see traditionalists reaching for their pens to condemn me for this view but I deal with the facts of life; wood, like 'Homo sapiens', wears out. We do not last forever, so how can you expect old furniture or the table that you purchased just a few years ago to look as it did when you first saw it in the furniture showroom!

A few decades ago, when waxing and french polishing was all that was available to the woodworking industry, stripping these surfaces was a very easy and straightforward job to undertake, but in recent years the introduction of finishing products such as cellulose, polyurethanes, polyesters, acid-catalysed finishes with their highly polymerised construction, or modern lacquers as they are termed, has brought new problems to the wood finishing trades.

First of all shellac finishes, such as french polish, together with wax and oil finishes are all termed reversible finishes. This means that the solvent used in the construction of the product, reacts to soften the hardened finished film, thereby permitting its removal. For example, in the case of french polish the solvent is methylated spirits. This solvent will reverse a hard dry film of french polish to its soluble form so that the wood finisher can completely remove it. Likewise, turpentine or white spirit will reverse a wax or oil finish, while cellulose thinners will reverse a cellulose surface film. The softened substance can then be wiped away with paper towels, waste wool or coarse wire wool, leaving the substrate bare and ready for re-polishing or finishing.

The problem comes with modern surface coatings which have been applied either by hand, spray

Stripping and Cleaning

Fig 18 Badly damaged top surface of an oak bureau before stripping.

or machine. These finishes, using lacquers with acid catalysts as hardeners, are extremely difficult to remove due to their make-up in manufacture; once hardened, the original solvents will have no effect at all upon the surface film. These are called non-reversible finishes. At this stage we must turn to specialist manufacturers who produce stripping compounds, mainly as pastes and fluids. Most of these contain methylene chloride, a non-caustic chemical which is a toxic substance and must always be used in an area with adequate ventilation. Wood finishers employing modern wood finishing techniques tend not to use traditional reversible solvents in removing modern surface coatings as they are weak and consequently it would take a great deal of time to finally remove a film with them. Instead, they turn to proprietary brands of wood stripping compounds, of which there are many upon the market today. Some are better than others. Some contain ammonia, some caustic soda, but many are non-caustic and contain, as previously mentioned, methylene chloride. These products vary in manufacture and the more powerful types will remove cellulose and synthetic lacquers, emulsions, polyurethanes, bitumen and stoving enamels.

Basically, there are two types of chemical stripping compounds: the fluid type and the paste type. The fluid type is ideal for all horizontal surfaces and mouldings, since the fluid can run into the fine corners of the moulding, whilst the paste types are better for stubborn surfaces as they can lie on a surface longer than a fluid, which means that they can penetrate the surface film better. The paste type strippers are also better for overhead surfaces and vertical surfaces, such as car body work, as they will not run off a surface.

At this point we must deal with the actual working technique of using a chemical stripping compound, for no matter how good a stripping compound is, the actual working technique is vital to the removal of the surface film; one is of no use without the other.

Method of removing a hard surface film

Prepare the item for stripping in an area with good ventilation as follows:

1. Place furniture or joinery on old dust sheets or paper to collect the soft deposits of the dissolved surface film.

2. Using a 'grass brush', which is a brush made of tampico and other vegetable fibres that are resistant to attack from strong chemicals, apply the chemical stripper liberally to the hard surface film and leave for at least 15 minutes. If the film is thick, this could take much longer and may even require a further application of the stripping compound. The basis of good stripping is to never let the compound dry out but to keep it moist upon the surface.

3. When you think that the surface has become soft or soluble, take a stripping knife or blunt cabinet scraper (with rounded corners) and scrape off the now softened surface film. Wipe off the waste with paper towels and deposit into a metal container. This process is rather messy.

4. Using either white spirit or cellulose thinners (preferably the latter), remove most of the remaining deposits from the surface by scrubbing with No. 4 wire wool.

5. Finally, using waste wool and a little of the cellulose thinners, wipe clean the remaining deposits from the surface of the now clean substrate.

6. It is at this stage that some wood finishers make a grave mistake; they think that as all the old surface has been removed from the wood that they can now start the re-finishing process. Nothing could be further from the truth. Wood has a grain and in that grain are deposits of the chemical strippers, which must be removed or they could damage, as well as create serious problems with, the re-finishing of the item. The way to remove these deposits is to apply a thin layer of the chemical stripper to the surface and leave for five minutes. Then, drag a soft brass wire brush along it in the direction of the grain and you will see that this releases most of the 'gunge' from the grain of the wood. Next, use clean cellulose thinners or white spirit with clean cotton waste wool and continue to wash out until there are no further traces of deposits left upon the wood.

7. One further process must be carried out to make sure that all traces of the old surface coating, combined with the chemical stripping compound, are finally removed from the substrate. First, allow the wood to dry thoroughly and then apply a sanding abrasive paper, such as aluminium oxide 150 grade, with the grain and dust off. Wipe the surface using clean warm water with just a spot of detergent and a soft cloth. A soft brush may help with this process, but whatever you use it must be done quickly. Cleaning the surface in this way does two things: firstly, it washes out any remaining unwanted deposits from the surface of the wood and, secondly, it will help to raise the grain slightly, which will push out any unwanted deposit held within it. Allow to dry out.

8. Sand down using a fine abrasive paper, such as 240 grade garnet, to obtain a smooth finish. The substrate is now ready for filling, re-staining and finishing.

Stripping and Cleaning

Having followed this procedure, you can rest assured that all traces of old surface films and strippers have been removed from the wood and that no matter what surface coating is to be used there will be no reaction.

Note: If the substrate is of veneer and has been glued with animal glues, then the use of water is not advisable. White spirit or methylated spirits (mineral spirit or alcohol) can be used instead.

I cannot stress enough the importance of removing all chemical waste from a stripped surface as most surface faults made in the re-finishing process are attributed to this point, and many wasted hours can be saved by following the procedure outlined above.

There are, of course, other ways of removing surface coating films other than by the use of chemical strippers. Polyester thick films, for example, are very difficult to remove when using chemical stripping compounds. However, the heat applied from a domestic hot iron (flat type) will easily break up the thick hard film, which can then be peeled off using a flat stripping knife whilst the surface area is still hot.

A cabinet scraper is of paramount importance in stripping antique surface films such as french polish or varnish, and it has the advantage that no harm can come to the surrounding areas during stripping, which is always possible when using chemicals. A cabinet scraper can also be used to strip a veneered surface, provided that the scraper itself is extremely sharp; a blunt scraper may damage the veneer.

(See Figs 19 to 22.)

Fig 19 The damaged surface of a mahogany table top before stripping

Fig 20 Note the slight curve pressure of the cabinet scraper

Fig 21 Hand dry stripping by cabinet scraper used in the direction of the grain

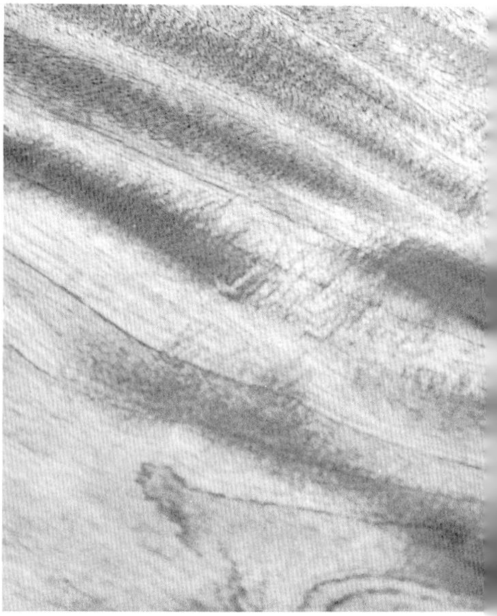

Fig 22 The finished result of a mahogany table top after dry stripping

Paint or varnish can be removed very quickly using a blowlamp or gas blowtorch. However, this heat treatment should never be used near glass, such as on window frames, or on wood where the grain will show, as the naked flame tends to burn the wood and the carbon deposit which is left behind is difficult to remove afterwards.

Electric hot air guns can also be used to remove paint or varnish films. These are ideal where glass is present, but again, care must be taken to avoid burning or scorching the wood.

Sanding machines can be used to remove some surface coatings. For example, portable belt sanding machines and orbital sanding machines can be used on soft surface films, such as french polish. Where painted surfaces are concerned rotary sanding machines are ideal since cross grain markings cannot be seen once the surface is re-painted.

In the commercial world of stripping surface coatings, hot caustic soda baths are used to remove paint, varnish, etc. This method is used extensively on pine woodwork, such as doors and cheap furniture, which have through the years accumulated many coats of paint. Commercial pine strippers use large baths which hold many gallons of caustic soda (sodium hydroxide), known as 'lye' in America. The chemical is a very strong alkali and to speed up the stripping process the baths are heated. Stripping by this method is quick. The item of woodwork is immersed in the tank and completely covered by the hot caustic soda (the mixture being 2 oz to 1 gallon/5 litres of water). It will take about one hour for the paint, etc, to break up and during this time the piece must be agitated within the bath. When all the paint has been reacted upon the whole piece must then be given a thorough wash using a power water jet in order to remove all traces of the chemicals and residues. The piece should then be allowed to dry out.

Stripping and Cleaning

Great care should be taken when selecting pieces to be stripped using this type of commercial undertaking. Under no circumstances should a quality piece of furniture be stripped in this way as the damage caused can be irreparable. I have seen furniture which has been immersed in a hot caustic bath where all the glues have disappeared and the fibres of the wood have been so raised in the process that many coatings of base or sanding sealers have had to be applied to restore the wood to a reasonable condition. A great deal of time and money has to be spent on restoring the wood before any kind of re-finishing can take place. However, for stripping paint, emulsions and varnishes off softwoods and cheaper furniture, this process is ideal.

Wood can also be stripped by using shot blasting guns. The Airless Spray systems use either glass, sand or other abrasive materials. These guns operate under high psi and should be used with great care and the operator should observe all health and safety precautions.

On some very old pieces of furniture, there are occasions when none of the above compounds will have an impact. Some of them were finished with a pigment (red lead) mixed in a medium of skim milk which, with the passage of time, forms an extremely hard finish. The only remedy in this case is to use strong ammonia (.880 type), which is very unpleasant to work with and it is advisable to do this job outside on a windy day. (See the safety check list at the end of the chapter.)

Not all surfaces need to be stripped. Antiques, for example, are only stripped if the surface is beyond saving; normally they are simply cleaned, but care must be taken when doing this not to damage in any way the original finish and patina which has developed over many years.

Whatever the surface coating, whether it be french polish, one of the new water-borne finishes, or synthetic lacquers, sooner or later it will need cleaning, and there are an ever increasing assortment of waxes, creams, emulsions, oils and aerosols on the market today which can be used for this job. Some are better than others and it is simply a matter of trial and error to find one which suits your purpose best. You can, however, make your own cleaning fluid, an emulsion 'reviver', which will clean off the residue of waxes or aerosols from the top surface without in any way breaking into or damaging the original surface or patina.

Here are three formulae which can be used for different surfaces:

1. *For cleaning french polished surfaces*
   1 litre of soft or distilled warm water into which has been dissolved one teaspoon of soft soap
   1 litre of paraffin (kerosene)
   1 litre raw linseed oil
   1 litre methylated spirits (alcohol)
   Add a few drops of eucalyptus oil for a pleasant odour

   Shake the contents and the result is a milky emulsion. It is necessary to shake the mixture constantly during use.

   Apply the cleaner to a rag or waste wool piece and rub onto the grimy surface. After all the dirt has been removed allow the surface time to dry and then finish off with a wax polish.

2. *For cleaning extremely dirty surfaces* (either shellac, varnish or synthetic lacquer finishes)
   1 litre of raw linseed oil

1 litre of turpentine
1 litre of methylated spirits (alcohol)
1 litre of white vinegar (mild acetic acid)
1 litre of soft water (either rainwater or distilled)
Add to the total volume 25g of fine pumice powder or rottenstone powder

Shake up the contents and apply as above.

3. *For cleaning panelling, doors, window sills, staircases, etc*

    Make up the following by volume:
    1 part linseed oil (raw)
    1 part white spirit (mineral spirit)
    1 part soft water (rain or distilled)
    1 part white vinegar
    1 part methylated finish – the clear spirit type (clear alcohol)

    Shake up the contents and apply to woodwork. Leave for a few minutes and then rub hard to remove dirt and grime. Allow to dry and finish off with either an oil or wax polish.

A few popular stripping compounds

There are countless brands on the market but listed below are a few of the most popular:

Nitromors Craftsman Original (fluid type) made by Wilcot Products Ltd
Nitromors Water Washable (paste type) made by Wilcot Products Ltd
Cyclone Stripper made by Cyclone Stripper Ltd
Strypit Paint and Varnish Remover made by Rustins Ltd
Caustic soda supplied by hardware shops, D.I.Y. stores, builders merchants, etc., or any chemical stripper containing methylene chloride.

Health and safety checklist in dealing with stripping compounds

1. Ensure good ventilation in working area.

2. Do not smoke or consume drink or food in the working area.

3. Always wear a face mask, goggles and proper clothing, e.g. gloves, etc.

4. Never allow the container tins to stand in sunlight.

5. Always unscrew the lid of container tins carefully to let out the pressure.

6. Never allow children or pets in the area where stripping is taking place.

Stripping and Cleaning

Surface Cleaner (Rustins)
This product is a blend of powerful solvents which will remove accumulated wax and dirt from furniture without harming the original finish. It should be applied with a soft cloth or, if the wood is very dirty, with 000 Steel Wool wiping off the solution of dirt formed with clean rag or paper kitchen towels.

Finish Reviver (Rustins)
If the original finish has lost its lustre or has fine surface scratches the Finish Reviver will restore the gloss and remove all minor surface defects. Water marks and heat marks can also be removed unless they have penetrated right through the finish to the wood.

Figs 23–24   Two popular branded products used for cleaning and reviving finishes

7. After stripping, remove all swabs and residue waste and incinerate in a suitable area.

8. When using a blowlamp or gas torch have a bucket of sand or a fire extinguisher close at hand.

9. If using a caustic soda tank always cover the top with a strong and heavy lid after use.

10. Always add caustic soda crystals to water, not the other way round.

11. If using a hot iron for removing polyester, do so in a well ventilated area and wear a face mask.

12. Personal hygiene is of paramount importance when using stripping compounds; protect hands with barrier cream and always wash them thoroughly when finished.

13. Keep containers in a metal cabinet.

CHAPTER SEVEN

Oils, Waxes and Lubricants Solvents and Thinners

Applying an oil to wood has been one of the simplest methods of preserving and beautifying wood for centuries. The American Indians, Greeks, Persians and Romans all used oils and fats for preserving the items they made from their wood and leather.

When wood such as oak was first used in the construction of English feudal castles and great houses, they did not attempt to give wood any finish – the wood simply aged and seasoned in situ with the passage of time.

During the early 15th century the new invention of this period was the 'fireplace' – built into the house or fitted as an extra. Take, for example, the Tudors: they had built within their grandest houses, huge fireplaces, burning wood logs and peat which gave off great amounts of smoke and dust. Some houses, such as the longhouses, had no such fireplaces, and lit fires in the centre of the living room giving off even more smoke and dust as there was no flue through which it could escape. The small amount of furniture that was made for the wealthy in those days was built mainly from oak remnants from buildings and soon became covered in soot and grime, and this was oiled in an attempt to clean it. The result is that the furniture of the period has a lovely patina which is really a build-up from the polishing and preserving by oils or fats for many hundreds of years.

The method was very easy to undertake. It consisted simply of rubbing the oils, which were mainly vegetable and animal oils, into the fibres of wood and these acted as a preservative and a cosmetic cleaning agent all in one go. A method which has since been rediscovered and copied as a new fashion preference in more recent times.

The traditional use of oils during the last century was mainly for table tops, bar tops, counters etc, but this was later superceded by better finishes such as french polish, cellulose lacquers, oil varnishes and now, water-borne finishes (acrylics). Oil polishing, however, is still used to this day, thanks to new improved blends of oils. The trouble with traditional oils, such as raw linseed or boiled oil mixed with turpentine, is that they take so long to actually dry out. Another disadvantage is that they collect dust and become grimy, although they do have the advantage in that the application of more oil actually cleans off this dust and grime, leaving the surface just as clean as when it was first applied.

In the practical treatment of a wood, simply apply the oil all over the substrate, leave for a while and then rub off until the surface is dry, then repeat the process at intervals until you have achieved the effect you want. The method is only used on bare wood and it should never be undertaken upon a sealed surface as the oil cannot penetrate the wood. Warming the oil before application is an added help for penetration into the fibres.

Modern methods of oiling do not differ in any way from the traditional – it is the oil that has been improved out of all recognition. The oils now used, such as tung oil and dehydrated castor oil, have drying constituents in them. The purpose of the modern oils on the market today is to aim for ease of application, quick drying and weather resistant qualities.

Recent modern wood finishing oils

Teak Oil

This was one of the first new oils to come onto the British market during the '60's, and unfortunately is misnamed. It is not oil of teak, but a man-made oil made up from a mixture of vegetable and mineral oils. It is still used to this day, and it has many qualities, but it must be remembered that it is for use on bare wood, not woods which have been surface coated.

Danish Oil

This oil, manufactured in the UK by Rustins, is a superb oil which is ideal for all hardwoods such as teak, oak, cedar etc. It is a special formulation based on tung oil (sometimes known as Chinese wood oil) and is extracted from nuts grown in China and South America. When processed, the oil is blended with synthetic resins which improve the drying hardness which is not possible with more traditional oils. Dryers and solvents are also added so that the viscosity is such that it can be applied either by brush, rag or spray gun.

This oil is not, however, a varnish and must not be classed as such. It dries within 4–5 hours depending upon climatic conditions, producing a dry, hard, non-sticky finish and, when dry, it can be burnished by using 0000 steel wool to produce a fine lustre. It can also be used as a primer or sealer for such items as pine wood furniture before applying other finishes. Danish oil can be used on interior or exterior woodwork such as gates, garden furniture, conservatories, greenhouses, cedar or pine cladding, doors and all kinds of turned items such as bowls and turned rails, etc – the list is endless.

The method of application is as easy as any other traditional oil – simply apply liberally with a clean brush or rag, wipe off the surplus and leave to dry hard. Then, with a fine abrasive paper, say 240 grit garnet – remove all nibs, dust down and apply a further coating. Do not apply it like a varnish, but simply wipe off any surplus 'lake' of oil and leave to harden. When dry this finish can be lightly burnished to produce a fine smooth silky surface which is water resistant, and will not chip or craze such as a varnish finish might. The use of this oil is ideal on such woods as iroko, oak or cedar, to name only three. Woods of this nature resist other finishes such as varnish and paint. Danish oil brings out the beauty of the natural colour of the wood yet does not leave a finishing film.

Scandinavian Oil Finish (Clear)

Made by Benjamin Moore & Co, USA. This is an interior wood finish which is a caster-resin ester penetrating type finish that toughens the fibres of the wood leaving a minimum of surface coating and giving a mellow sheen. (Approved by the US Dept of Agriculture.)

A glossary of oils used in traditional and modern wood finishing

Linseed Oil   Raw linseed oil is obtainable from the flax plant seeds, grown in such countries as UK, India, U.S.A. and Argentina. It is a light yellowish transparent oil used as a lubricant in french polishing and as a medium for manufactured trade oils and varnishes and, in addition to these, is a most important oil as it has the qualities of film forming when used in the process of making alkyd resins. In its pure state, however, it is slow drying and has poor water resistance.

Boiled Oil   This is linseed oil which is heated with the addition of drying agents. It has better water resistance but darkens further with age and becomes brittle. It is added to paints in manufacture due to its faster drying film forming qualities.

Soya Bean Oil   A pale yellow slow drying oil, which does not darken with age. Used in the manufacture of pale and non-yellowing paints.

Tung or China Wood Oil (originates from China)   A slow drying oil in its normal state but very resistant to alkalis and water. Used in the manufacture of Danish Oil, and in combination with resins for use in water resisting varnishes.

Castor Oil   A vegetable oil extracted from the castor oil plant or 'Palm Christi' plant. The oil in its natural state never dries out. It is converted into a drying oil by dehydration and can then be used as a plasticizer or for further formulations with resins. If added to nitro-cellulose lacquer, it will slow down the drying time which is an advantage with 'brush on' applied lacquers.

Poppy Oil   This is used as a lubricant in the french polishing technique as an alternative to linseed oil.

Mineral Oils   These are oils termed 'technical white oils' which are light, transparent oils, and are manufactured from crude oil. They can be used as a lubricant for the french polishing technique, or as plasticizers for use in fillers, varnishes and lacquers. One of the main advantages is that these oils can be used with safety for the oiling of cedar shingles on house roofing, and will not promote mildew.

Rape Seed Oil   Those marvellous bright yellow fields of rape flowers – the seeds are processed to produce oil for cooking and other uses, and can be used in lubrication in the french polishing technique.

Creosote Oil   A product processed from coal which is ideal for its preserving qualities, and frequently used on such items as power poles, telegraph poles, fences, external woodwork etc. It must not be used anywhere near foliage, children, pets, animals etc.

Manufactured Mineral Oils for Exotic Woods   There are various such oils on the market, both clear and coloured – all shades including green. My pet hate is the red lead coloured variety for cedar and teak. This completely destroys the beautiful natural colour of these woods and looks ghastly after weathering.

Teak Oil   A manufactured oil using a blend of vegetable and mineral oils and driers. Used for both interior and exterior work for all bare-surfaced softwoods and hardwoods and for the general cleaning of furniture.

Danish Oil   A manufactured oil using a combination of Tung oil and synthetic resins, which will dry bone hard and which is used for all woods internally and externally, and for all kinds of manufactured wooden items.

Turpentine Oil (classed as genuine turpentine)   Turpentine is a vegetable oil that dries by oxidization. There are two basic types: a) Pure gum turpentine, which is made from the distillation from pitchy pine stumps, while b) Gum turpentine, is superior as this is in its natural state and collected as it exudes from the tree, rather like the collection of rubber latex, and then processed. It is costly, mainly comes from the USA, and it has a lovely scenty aroma. Turpentine absorbs oxygen, thus forming a film, and is therefore used in the manufacture of quality paints and varnishes, wax polishes and creams, and is also used in the manufacture of medicinal products such as in creams, ointments etc.

Turpentine Substitute or White Spirit   This misnamed oil is a mineral oil and is manufactured basically as a solvent for oil stains, paints and varnishes and for cleaning oil-paint brushes. It is also a solvent for wood fillers and for stains, etc, and is a very inexpensive product. The oil, however, does not oxidize and therefore is not to be used in good quality paints or varnishes as a thinning agent. (In the US this is known as mineral spirit).
Note: The word spirit here must not be confused with methylated spirits.

Camphor Oil   This pleasantly scented oil is used in some furniture revivers and is a general oil for the removal of stale odours in older furniture. When mixed with a small amount of olive oil, it can be used for the removal of slight ring watermarks on french polished surfaces. (Formula – 5 parts olive oil to 2 parts camphor oil by volume.)

Paraffin Oil (Kerosene)   These oils come as coloured transparent liquids depending upon which oil company produces them, and they are used in the make-up of furniture revivers or for cleaning down greasy surfaces prior to repainting or staining. They are much cheaper than white spirit and are ideal for treating woodworm infestation when mixed with a little creosote oil or acetic acid. (Formula – to 5 litres of paraffin oil add ½ litre of creosote oil or acetic acid.) Apply liberally to the infested wood., There is a strong odour with this mixture and items of furniture or joinery so treated should be placed in a well ventilated area until dry.

Solvents Solvents are the transporters of solids – they exist mainly for the application of the solids, of lacquers or paint and nothing more. The solvent simply dissolves a solid leaving it at the stage when, upon evaporation, it becomes a film. The solid in this meaning is the actual chemical body or organic make-up ingredients of the surface coating. There is a complication in that one solvent can be dissolved in another and the resultant solvent can then be used as a solution. This is classed as a mixed solvent, or the part of the solvent can also be a latent solvent which comes into action under certain chemical conditions, whilst being part of the whole. The miscibility of solvents are very complex and one needs to be an industrial chemist to understand them.

However, to refer to a solvent means that the term must be used in relationship to its particular group of materials. For example, turpentine may be a solvent for an oil varnish, but is of no use as a solvent for a cellulose lacquer.

Thinners It is here that the word 'thinners' becomes confused with the word 'solvent'. In fact, the thinner is a mixture of solvents simply to reduce the viscosity (to thin down a lacquer, varnish, paint, stain etc) of a basic manufactured surface coating, but the thinners must be compatible with the solvents to some extent, and when used must not upset the true balance of the original solvent and solid make-up ingredients resulting in the surface coating film. It must be emphasized that the correct thinners must be used when dealing with cellulose, acid catalysed and polyurethane lacquers: to upset the balance of the solvents in these products can lead to the complete break-down of the lacquer film. It is therefore necessary to use only the correct thinners as formulated by the product being used. (Cellulose thinners, precatalysed thinners, acid catalysed thinners, polyurethane thinners, white spirit thinners, etc.)

Most surface film faults, such as blooming (foggy formation), and non-drying patching of the lacquer film, are produced by the use of incorrect thinners. It must be emphasized, therefore, that only the thinners from the original manufactured source can be used. Do not be tempted to add like-thinners of mixed manufactured sources together because the products may not be compatible. In other words, cellulose thinners of one brand may be different to cellulose thinners of another brand.

Lubricant Oils These are widely used in wood finishing methods, mainly in the application stages. In the french polishing application, linseed oil (raw), is a very common oil, while poppy oil, rape seed oil, soya oil and white oil all have one common aim in the production of a smooth fine finish, which is, for the shellac-filled rubber to slide on or over the substrate. Most of these oils, however, are miscible and get buried within the film of shellac polish. While linseed oil can become embedded in the film of shellac, white oil can act as a plasticizer and can cause the polish to become soft. Also, when the oil oxidizes, it can break up the surface film, causing crazing. It is a matter of choice for the wood finisher to chose which he or she finds better for the job in hand. It is always advisable when french polishing to warm the chosen oil as this helps in the smooth running of the rubber with shellac based french polish. (To warm the oil, place a quantity in a small container and put this into a larger container of very hot water. Never heat it up using a naked flame.)

White spirit is used as a lubricant with wet and dry silicon carbide abrasive papers in modern lacquer finishing during the flatting process. Water can also be used with a little detergent or plain soap as a lubricant on these cellulose finishes. When using water, it is better to use either rain water, which is softer, or distilled water, as certain chemicals in tap water in some hard-water areas can cause problems when using acid catalysed lacquers.

Beeswax   The wood finishing trade depends entirely on the by-product of two insects – both female. One is the lac scale beetle, the '*Laccifer lacca*', the other is the bee, which provides beeswax. This substance is one of the little miracles of nature for which there is no real synthetic substitute. Beeswax is an animal wax produced entirely by the female worker bee. The wax is non-toxic and is harmless to human skin, and is one of the most important waxes used by the wood finisher in various composite forms as well as in its natural state. It is a solid wax, fairly hard and with a texture like soap.

The bee secretes the wax which she uses to make combs and sealers for the comb ends, called 'cappings', which prevents the honey from running out to waste. When the honey is harvested the combs and cappings are melted and separated from the honey by the bee keeper, who has, therefore, two cash crops – honey and beeswax. The melting point of beeswax is 140°F, and raw beeswax is strained, cleaned and refined, and then poured into clean moulds and allowed to set. The best beeswax is a fine, deep buttercup to primrose colour with a good aroma.

In the natural state, beeswax has been used for hundreds of years and the early makers of furniture were quick to use the wax for polishing wood and leather. Broadwoods, the English piano manufacturers, were polishing their harpsichords and pianos up to 1815 with beeswax and turpentine before turning to french polishing.

Today, beeswax is used extensively throughout the antique trade and on some modern quality reproduction furniture in modified form. In addition, the wax is used as a stopper for small cracks, chips, holes and blemishes on shellac surface work. Beeswax can also be obtained in bleached form to produce a neutral colour for use on pale coloured timbers such as pine, but the bleaching process makes the wax harder and less supple to use. In this state it is sometimes called white beeswax.

Many furniture waxes and creams use beeswax as a part ingredient, and it is of interest to note that the wax is also widely used in industry and trades as varied as sailmaking, car manufacture (sprayed inside door frames to prevent corrosion), and in pharmaceutical products such as creams. It is also used in the manufacture of traditional church candles and high quality domestic candles.

Whilst beeswax polish has become very expensive in today's market place, it can be made by the amateur very easily.

Formula: To produce a 'home-made' furniture wax polish, take a medium-sized pan and into it place an ampty 1 3/4 lb tin (or similar). Pour water into the pan until it comes about half-way up the tin and bring to the boil. Turn the heat down so that the water continues to boil gently, and into the inner tin place about ¼ lb of beeswax and allow it to become liquid. Slowly and carefully add about half a cup of pure (not subs) turpentine, and stir in with a wooden spoon to mix well. Then add two teaspoons of very strong ammonia (.880 type, not the household type), and about a teaspoon of either fine pumice powder or rottenstone, both of which are abrasive materials which will give the polish just that little 'bite'. The addition of a little carnauba wax will make the polish harder and will produce an easier shine. Mix together thoroughly with your wooden spoon, then turn off the heat and allow to cool. When set take a small sample and try polishing a piece of wood. If the polish is stiff, add a little extra turpentine and ammonia; if too runny, add a little more beeswax, and so on until you have the right consistency – in either case re-heat and mix in. You could also add a little colour (say brown umber), or a little perfume such as camphor oil. When you are satisfied with the mixture, pour it off using a sieve into a clean tin with a lid and leave to harden off.

Note: Remember that all the materials that you are using are flammable so take care during the mixing process. It is far safer to do this process outdoors.

Carnauba Wax   Another very important wax used by the wood finisher. It is yellowish, non-toxic and harmless to the skin, but differs from beeswax in many ways. First of all it is a vegetable wax – a natural exudation gathered mainly from the Brazilian palm tree at a rate of approximately 6 ozs per tree per year. Secondly, it is a very hard wax – so hard that you need to break it into smaller pieces by using a hammer – and with a melting point of 185°F, it is without doubt one of the hardest waxes. It is ideal for use on the turner's lathe for final finishing, for spindlework for instance, by using a fast lathe speed and holding the wax against the work lightly but firmly, the friction melts the wax. This can be followed by 240 garnet papers to provide a super-fine grain-filled hard wax finish. It can also be melted and used as a stopper for large areas such as around knots in pine, and can be incorporated with other waxes such as standard furniture waxes, burnishing pastes and creams, which need to be harder and less sticky for cellulose and synthetic finishes – for instance, it is used in most car polishes for lacquer surfaces. Carnauba is supplied in two grades: fatty grey and prime yellow, which is the better quality.

Japan Wax   Originating in Japan, this is not really a true wax like beeswax or carnauba, but a blend of vegetable fats with other products such as pigments or shellac. When blended with shellac, it makes an excellent stopper which does not shrink, is hard and takes french polish or lacquers over it. A 'shellac stick' is made in this way and is the same size and shape as a piece of sealing wax. It comes in a variety of colours – white, black, brown, red, cream etc, so that scratches or imperfections can be filled with the same colour as the surrounding timber. It must be used with a very hot knife to melt the stick – never a match which will simply cause carbonation and blacken it. In modern finishing techniques, a cellulose wax is classed as a Japan wax and is used as a stopper in the same way. Shellac sticks come in boxes of assorted colours and are ideal for the journeyman-polisher where quick repairs can be made to small faults in furniture on site.

Beaumontage Wax   This is a fairly soft pigmented wax which can be used as a stopper for shellac or cellulose finishes. It is made up of various waxes, both natural and mineral.

Paraffin Wax   This is a mineral wax derived from the petrochemical industry, made from paraffin oil – either white or clear, and it is commonly used throughout the trade. It is a fairly solid, soft white clear wax and it is used mostly in combination with other waxes to produce a soft furniture polish. The fact that the wax has 'non-slip' properties means that it is frequently used in floor polishes, and, last but not least, it is cheaper than beeswax polishes. It is easy to identify a cheap wax polish by the smell – it has a paraffin odour.

Wooden floors and linoleum are still best polished by wax particularly when a modern electric floor polisher is available to ease the task. It can also be used to ease sticking drawers by rubbing onto the bottom slide rails.

The following waxes are also used in the manufacture of various polishes, frequently as substitutes for beeswax and carnauba waxes, and used in combination with them as a Japan wax:

Candelilla Wax   A vegetable-type wax originating from a Mexican plant with a melting point of 150°F.

Ceresin Wax   A hard wax which is man-made from a blend of paraffin wax and other synthetic waxes such as ozokerite.

Chinese Wax   A wax which insects secrete on to the twigs of Chinese ash trees.

Bleached Montan Wax   Another man-made product manufactured from brown coal or peat and used as a substitute for carnauba.

Ozokerite Wax   This sounds Russian, but it is a petro-chemical extract and is used for blending with other waxes.

Lac Wax or Shellac Wax   Lac wax is basically a hard wax similar to carnauba wax, and it is a by-product produced during the process of bleaching and de-waxing shellac. It is mainly used by various manufacturers in the production of shoe polishes and thin furniture creams. As far as wood and leather are concerned, lac wax is used at times as a substitute for carnauba, candelilla or montan waxes, but unlike carnauba, lac wax is not consistent in quality, and varies from different growing areas of India.

Silicone Waxes   These are special kinds of waxes which have outstanding water repellancy and hardness. These waxes are not true waxes in the ordinary sense of the word, but both natural and synthetic waxes to which have been added chemicals called 'silicones'. They are complex in their production and there are a great variety available on the market today in paste, cream and aerosol forms. They can, however, create many problems for the wood finisher, as they 'plant' a secondary surface onto the substrate, which builds up with the passage of time collecting dust and grime and can eat into a surface coating. Great care is needed in the application of these waxes to avoid trapping in grease and dust onto a surface. Aerosol spray polishes are acceptable for very large boardroom-type table tops, as, unlike solid waxes, they do not smear, but they must be used with care.

Wax Finishes   A wax finish is far longer lasting than an oil finish. This is due to the fact that a wax

Oils, Waxes and Lubricants, Solvents and Thinners

finish leaves a thin film deposit upon the surface of the wood, and, unlike oil, leaves no greasy deposit for dust and grime to cling to the surface. A wax finish also enhances the beauty of wood as well as leaving a dry film which reduces wear and abrasion. If silicones are added to the wax, then the surface becomes slightly resistant against water and heat.

What waxing entails is simply rubbing small amount of workable solid wax onto a substrate using a piece of mutton cloth, and burnishing until all traces of smears are removed.

Polishing waxes are available in either paste or liquid form, and the latter also incorporates the aerosol can form.

Basic wax polish, either for wood, for furniture or for car bodies contains not just beeswax, but also carnauba wax, resulting in a harder finish and brighter lustre or shine. Some wax polishes contain turpentine for furniture use, or paraffin oil for its non-slip properties if used for floors. The use of these solvents in wax polishes makes them easier to apply or spread, and in the buffing action they evaporate leaving the wax deposit behind.

Figs 25–26   Waxing an antique chair

A glossary of waxes available

Transparent or colourless wax for antique furniture, non-silicone paste.
Antique wax polish, non-silicone paste in brown/black colour.
Antique wax polish, non-silicone paste in black colour.
Wax polish in light brown colour, non-silicone paste.
Liming wax polish (white colour) – ideal for the liming effect on wood.
Floor wax polish in a white transparent colour (paraffin wax).

Floor wax polish in a light brown colour (paraffin wax).

For the waxing of pine furniture and pine panelling there are a number of special non-greasy waxes available, such as warm pine wax polish, cold pine wax polish, antique pine wax polish, and these are specially formulated for pine finishing and easier to apply than standard basic wax polishes. Car body wax polishes contain various waxes and abrasive materials such as rottonstone, sometimes known as Tripoli Powder.

Modern Emulsion Wax Polishes

These are an emulsion blend of aniline dyes and lead-free pigments and solvents, etc and are very easy to apply and come in a vast range of colours.

Aerosol Wax Polishes

Most of these contain beeswax and silicones and they are popular due to the ease of application. They are an excellent form for applying wax to very large areas or where standard wax paste is difficult to apply.

It is most important to use the correct type of cloth and in my opinion, there is none better for wax application than mutton cloth which is a pure cotton mesh cloth. Yellow dusters, stockinette cloth, old sheets etc, are of little use as they have no 'bite'.

Health and Safety check list when using oils, waxes and solvents

1. Wear plastic or rubber gloves.
2. Use a hand barrier cream.
3. Allow plenty of ventilation in area being worked.
4. All swabs of teak oil and Danish oil must be burned after use in a special outside incinerator as the fumes given off are toxic, and the substances can also be explosive. They must not be left in a container or waste bin or they could spontaneously combust.
5. Do not smoke or consume drink or food in area worked.
6. Wash hands with a plain soap after handling.
7. Use a hand cream after working.
8. Wear protective clothing – apron, etc.
9. Keep materials away from direct sunlight.

10. Keep children and pets away from odours.

11. Cellulose materials should be kept in metal containers and stored in a cool place.

CHAPTER EIGHT

Metal Finishing

Most furniture and joinery construction today has some part of it made of metal. Modern designs incorporate a great deal of metal by way of construction supports, legs, and, in some cases, items such as shelving are made completely of metal.

The finishing of metal is not just a cosmetic application and, unlike wood, it has a nasty habit of rusting; therefore in metal finishing the whole object must be:

1. To prepare the substrate by degreasing;

2. To prevent corrosion;

3. To apply a cosmetic resistant finish by hand or spray.

Before any type of finishing to a metal substrate can be commenced the metal has to be in a clean and stable condition and no grease of any kind must be present. The grease may come from the manufacturer or the metal may have some kind of grease or treatment to preserve it whilst in transit or in store; the essential preliminary treatment, therefore, is to produce a clean surface so that it will accept finishing coating films. This is achieved by degreasing the surface by such methods as shot-blasting, which can also remove rust, or by using a solvent degreaser which is now the more popular method. Shot blasting is only to be carried out under health and safety precautions and suitable clothing and safety equipment should be used by the operator. Solvent degreasing is usually carried out by using hydrocarbons or more powerful solvents such as carbon tetrachloride, and care should be taken with these substances to avoid inhalation of fumes.

These solvents, however, are not always efficient as they can sometimes leave a deposit themselves, such as when using a hydrocarbon solvent like white spirit.

Once the metal is free of grease, one of the better methods of preventing corrosion is to phosphate the surface *immediately* after degreasing. Zinc phosphate (although other metal phosphates can sometimes be used) is applied to the ferrous metal surface, resulting in a fine formulation of zinc being deposited upon the surface of the metal to provide a barrier against future corrosion, and this surface also provides a key for all subsequent surface coatings. On the market today are many such de-rusting fluids and pastes which contain phosphoric acid which is the basis of most metal primers. A description of those available is as follows:

Zinc Phosphate (metal primer)

This is an ideal hydrocarbon solvent metal primer. It is a non-toxic pigmented primer with an excellent rust inhibition and is also a high build finish, which is an advantage for flatting down using wet and dry silicon carbide abrasive papers 400–600 grit. Sometimes red lead is added to the colour but this does vary between manufacturers. Oil based pigmented colours can be applied over this primer when completely dry. This method is ideal for all exterior steel structural work but is not suitable for metals which are to be immersed under water. The primer can be applied in this instance by brush using white spirit as thinners.

Zinc Chromate Metal Primer

This is a metal primer which is yellow in colour and is often called a universal primer. It is rust inhibitive and has a good flow quality and is resistant to marine atmospheres. The primer can be applied by brush or spray gun. Thinning is by a hydrocarbon – white spirit. If the substrate is of a poor condition this primer is ideal to level the coatings due to the ease of flatting by abrasive papers.

This primer can be used on non-ferrous metals such as aluminium and light alloys., It is slow drying and the top coatings should be applied as soon as the primer has dried. Most pigmented decorative top coatings can therefore be applied.

Red Lead

This is one of the most popular and well-known of the metal priming coatings which is rust inhibitive. It is bright red in colour – as the name implies, but the quality is also based upon the amount of genuine red lead pigment contained within the solvent which is a hydrocarbon. The primer has an excellent high film build quality. One of the most popular uses of this primer is on heavy structural steel work, either applied by brush or airless spray systems.

There are, however, red lead primers which use a cellulose solvent and these are for use under all cellulose based surface coatings which can be sprayed on and these are mainly used in the car, auto, or allied finishing trades.

Cold galvanizing (zinc-rich primer)

This is a heavy, rich 90% pure zinc dust suspended within a medium of epoxy resin. The preparation of the substrate must be by blast cleaning to remove all grease and other impurities. The zinc particles when applied to bare metal will weld themselves together upon drying, which is very fast – approximately within thirty minutes. The finish can be left without any pigment colour top coat finish applied to it, and the great advantage of this primer is that it is water resistant. The solvents used in this product are complex and thinning must be in accordance with manufacturers' instructions. It can be applied either by brush or spray gun.

Etch primer

This is a system of pre-treating clean ferrous and non-ferrous metals to make sure that the subsequent coatings will adhere without surface film refusal. The primer is a catalysed material where the acid is actually phosphoric acid, and the thin film formed takes place upon a metal surface which is normally applied by spray gun to obtain the best effect, providing a 'key' or maximum adhesion for surface coatings to follow.

Some etch primers have a little colour added to provide the operator with evidence of coverage during use.

It is possible to have etch primers formulated for other purposes which will deposit a thin film of zinc as well as phosphating the substrate metal at the same time. The whole object of using this thin fluid primer, which is applied after degreasing the metal concened, is where direct phosphating has not taken place or is not practical. The reaction of this fluid on metal etches the surface of the metal at the same time as becoming a pre-treatment. The resultant finish is therefore corrosion free and also provides a good key for the second primer and cosmetic surface coating that follows. One point that must be made here is that unlike some of the forementioned metal primers, the etch primer itself has no long term protection unless overpainted or oversprayed with a normal metal primer, further followed by one or two top coatings of a suitable cosmetic surface coating.

Corrosion of metal

Unlike wood, metal corrodes, or, more commonly, rusts. Any car owner who lives on the coast knows full well the meaning of corrosion due to the extreme salt-laden atmosphere which accelerates the breaking up of the surface protecting film. Rust is the cancer of metal due to the reaction of water upon exposed metal areas which expands and breaks up the metal structure, reducing the solid metal into thin layers of rusted sheets, and finally into holes or areas where the corroded metal turns to metal dust and falls away.

Corrosion, which requires the presence of both air and water to take place, is the enemy of metal, and when Air Forces and major airlines of the world have temporarily unwanted aircraft, they are often parked in hot, dry deserts where the air is dry and unpolluted, and corrosion reduced to almost nil. When required, they are flown back into use with very little damage to the metal work, emphasizing the importance of anti-corrosion methods when dealing with very expensive machinery.

Metal forms a great part of our material possessions – central heating tanks, window frames, motor vehicles, lawn mowers – the list can be endless, and they require protection. It is a constant fight to slow down the process of rusting and, although it cannot be completely eradicated, by using current technology the process of decay can be greatly retarded.

Metal Finishing

A standard method of treating corrosion on ferrous metal using a proprietary rust remover, anti-corrosive paste

The following procedure could be applied to car body repairs:

1. Using a scraper, remove pockets of rust deposits.

2. Using a wire brush, remove further loose deposits and finish off with steel wool No. 4 grade.

3. Check that there are no grease or oil deposits by using a degreaser such as carbon tetrachloride.

4. Brush the rust remover paste liberally onto the rusted areas of the bare metal only, and rub or scour into the metal using steel wool. Leave on for 15 minutes and then apply a further coating of the paste and immediately wipe dry using a clean cloth, then leave for thirty minutes.

5. Sometimes, before applying fillers, the corrosion may be so severe that an actual gap or hole has been made in the metal, thus affording no key for the filler. The remedy here, if the perforation in the metal is no more than 25mm/1" in diameter, is to use a coarse fibreglass filler which is manufactured for bridging rust holes and splits. When this is dry it can be sanded down and a normal paste filler used over it to obtain a smooth finish. If the hole or gap in the metal is of an extra large size use a piece of thin metal perforated gauze, fixed by normal fillers, to bridge the gap in the metal, which is finally embedded using a normal catalysed paste filler.

6. In a situation where there are no holes present, apply a standard catalysed filler to the treated areas and flat down when hard using wet and dry silicon carbide abrasive papers.

7. If, by chance, the abrasive cuts into the filler and exposes the metal, this must be retreated with the rust remover paste and re-primed.

8. As soon as possible apply the first coating of a metal primer and allow to dry. This may require a second coat of primer. When dry flat down using wet and dry silicon carbide papers with water as lubricant, making sure that you have a smooth undented surface.

9. Wipe off all sludge or gunge and allow to dry.

10. At this stage it is best to let the primer coatings harden off for at least twenty-four hours in summer and longer in winter.

11. To obtain the best results, two surface coatings of an undercoat can be applied at intervals followed by two coatings of the final cosmetic finishing surface coatings to complete the process. This method can be carried out using either cellulose, synthetic or oil based paints – the procedure is the same.

Note: The paints must never be intermixed – keep to the medium of your choice from the start. The white spirit thinners for oil and thinners for cellulose are not compatible. Depending on which medium you decide to use –paint can be applied by brush while cellulose is better applied by a spray gun system. Cellulose in this instance means synthetic catalysed lacquers, as very little pure cellulose is used nowadays.

Proprietary rust remover, and preventer, contain phosphoric acid. Some are fluids, some pastes, but the action is the same. The thick paste is ideal for vertical surfaces and overhead awkward situations, and the fluid type is better suited for small items that require dipping. The choice of which type to use is a personal one according to the job in hand.

Allow three to four days for the final finish to harden after which time it can be burnished if required by using an abrasive burnishing paste or cream. A powered rotary wool pad is an added advantage for this final stage.

Dealing with metal items on furniture

There is always a problem when restoring an old piece of furniture which has metal catches, hinges, bands, struts, handles, screws, bolts, nails etc. These must be treated very carefully as they are a part of the whole of the furniture, and it is always a good plan to remove them if at all possible without damaging the woodwork.

Take, for example, a Jacobean oak chest which has blacksmith-made bands and hinges which are in a bad corrosive state. These are not always easy to remove and if this is the case, it is far better to leave them in position, although the rust still has to be dealt with. First of all it is a good plan to mask out the wood from around the metal fittings. The cleaning of the metal – in this case iron – should be carried out using steel wool and a brass wire brush, being careful not to scour the surrounding supporting wood.

When all is as clean as it is possible, give the ironwork a thin coating of a rust remover, either fluid or paste, and allow to dry. When dry, apply a thin coating of lubricating oil and wipe dry: this also applies to any screws that may have been removed which should first be dipped in a de-rusting fluid and then dried before re-fitting. Brasswork or copper requires a different method. Unlike many antique dealers, I like all brasswork to look clean on antique furniture as it would have been when the furniture was first made. This can be achieved by removing the brass items such as handles, castors etc and cleaning them using 0000 steel wool and a burnishing paste. The worst thing to do is to try and clean brasswork whilst it is still fixed to the furniture as this will show scouring marks on the wood. And neither should proprietary brass cleaners be used on fixed brasswork as they can leave unpleasant white deposits in the grain of the wood around the fittings. Remember too, that in addition to the more obvious items of brasswork, things like screwheads should also be cleaned before refitting. In my opinion, clean brasswork brings alive old furnitue and there is nothing which brings such joy to my eyes as an old chest of drawers with a lovely patina and shining brass handles!

Metal Finishing

Surface finishing coating of metal

Various finishes are available for applying a cosmetic or weather-resistant film to metal. After degreasing and treating for anti-corrosion, good preparation is essential if you wish to achieve a visually perfect final finish. It is important to use suitable primers and undercoating films which should be flatted down to remove any fault that could later show up when the final gloss coat is applied.

Finishing coatings of either gloss or satin can be applied by spray gun, electrostatic deposition or dipping, or in the case of structural steel, the finish may be hand brush coated using materials specially formulated for this purpose.

The materials used are mostly oil, bituminous, nitrocellulose or synthetic cold-curing based commercial finishing, including air-drying, based upon epoxies, alkyds and acrylics. There is, however, one other metal finishing surface coating film that has become popular in recent years – that of powder coating.

Thermosetting Powder surface coating

Thermosetting is a term used for setting permanently when heated. The plastic substance undergoes a chemical change which cannot be reversed or remelted, unlike a thermoplastic substance which can. This is a commercial system where metal surfaces, after normal degreasing and anti-rust treatment are sprayed using a resin polyester powder applied by electrostatic spray gun systems which form as a powder upon the substrate. This is then fused in a stoving oven at a temperature of 200°C for approximately ten to twenty minutes. The result is a perfect gloss or satin finish which can be smooth in appearance.

This high quality polyester powder coating can be applied to all metals including aluminium, and has a high resistance to weathering and has colour retention properties. A full range of colours are obtainable from specialist manufactures in this field. The equipment required for this method of application is an electro-static spray gun system, and a dry spray booth with special extraction for powder retention (from over-spray) and a stoving oven. One great advantage of this system is that there is no fire hazard due to the fact that there are no solvents used in the process and therefore outside the petroleum regulations of flammable liquids. The process is ideal for all component articles of awkward shapes which are difficult to surface spray using conventional spraying systems, and, due also to the lack of solvents, the overspray powder can be collected and re-used, which is not possible with fluid surface coating applied by spray gun.

The covering power of powder surface coating can produce a film thickness of 50 microns and this can cover an area of 12–15 square metres. (The average thickness of a coat of paint is approximately 25 microns).

Another method in the application of powder coating is the fluidised bed system, where the substrate is pre-heated and immersed in a tank containing powder which is thus fused and produces a much thicker coating film.

Health and safety check list in dealing with metal finishing materials

1. Wear suitable clothing, including a hard hat if working on a building site.
2. Wear eye shields and plastic or rubber gloves when handling acids, fillers etc.
3. Never smoke or consume or store food and drink in the work area.
4. Never leave open tins of materials near children or pets.
5. When using catalysed fillers and solvent finishes, have good ventilation.
6. During powder coating application, special precautions must be taken because of the dust hazard involved and face masks should be worn. Good extraction of fumes, etc, is essential when using these products.
7. When using steel wool, always use gloves to protect your hands.

UK commercial users must conform to the COSHH regulations (Control of Substances Hazardous to Health) 1988. US commercial users must conform to OSHA regulations (Occupational Safety & Health Administration). (Other countries will have similar regulatory bodies and legislation governing health and safety at work and readers in these areas should aquaint themselves with local conditions before commencing any work.)

CHAPTER NINE

Colouring Wood

The art of colouring wood has been practised for thousands of years. The ancient civilizations decorated their furniture with a variety of finishes including pigments, varnish and gold leaf. The practice continues today; but why is this so? Why do we colour wood? There are a number of answers to this question:

1. Tradition.

2. To give wood an artistic decorative appearance.

3. To give inferior woods, such as pine, the appearance of more expensive woods. This practice is used widely by major furniture manufacturers today.

4. To match new woods or veneers to old, such as in antique restoration work.

5. To preserve the wood; most colouring actually preserves the wood.

6. To emphasize the natural grain of the wood. In many cases this is best achieved through colouring.

To a large extent, tradition dictates the colour of our furniture today. We expect oak, for example, to look dark like the antique pieces discoloured by smoke and soot from open fires. When we think of mahogany, we think of a reddish wood; this dates back to the introduction of the first of the species from Honduras, which happened to be a red variety. When we think of walnut, we think of a wood which resembles the colour of the nut itself.

Manufacturers of furniture cater to our tastes, but as our tastes are influenced by the past it is hard to break away from traditional colouring. Happily, the situation is changing. We are now seeing such woods as ash coloured green, which shows off wonderfully the grain of this beautiful wood. Black is being used on oak together with a red liming colour, while mahogany is being coloured blue with white liming in the grain to create some wonderful effects. I, personally, have re-polished grand pianos using pink and green pigments to produce some very exciting results, and why not? Must we colour pianos and other furniture in mahogany and walnut colours forever? Surely not! Furniture, like all joinery finishings, can be exciting if only colour is used more openly and daringly. It may be correct to use traditional colours where antiques are concerned but most of us now live in a new age where music, painting, the theatre and arts in general all have something new to say, so why not let modern thinking and technology loose on our joinery and furniture.

Pigments and Stainers

Broadly speaking, a colour which is applied to wood is obtained from either organic or inorganic sources. Organic pigments are derived from natural earths, such as yellow ochre and burnt sienna. Inorganic pigments are made of metals which have been chemically treated to produce fine powders and they include such colours as titanium white (titanium dioxide) and antimony oxide, both of which are white. A list of pigments in common use is given in the table on p. 00.

Pigments are used to colour paints, emulsions, lacquers and varnishes. Some are strong in colour whilst others are weak. There are even some which are transparent. Some are added to a medium to alter the tone of a colour while some have strong properties of colour obliteration. Most of the pigments we use today are manufactured chemically but some of the organic ones have been around for many hundreds of years. The Renaissance painters, for example, used such pigments as yellow ochre, venetian red, sienna and lamp black. Pigments do vary in their stability and resistance to the action of natural light. Some, particularly the organic types, are prone to fading, while others actually darken or change their shade when in dark or shaded areas. In recent years, developments in the production of pigments have enabled manufacturers to produce not only brighter colours but also colours which are not affected by the action of light.

Pigments are also available in the form of stainers, which are very strong, thick concentrated pastes or fluids. They are supplied in tubes, like toothpaste tubes, for the ease of adding colour to paints, emulsions, etc, and are either water or solvent based.

Universal Concentrated Stainers are mixtures of complex metal pigments in solvents and they can be added to either shellac or cellulose finishes. They are frequently used for colour matching and are available in red, blue, green, yellow and black. They have good light fastness and, being much thinner than other stainers, can be used by spray gun or by hand finishing methods.

An important point to remember when dealing with pigments is that unlike stains they do not penetrate the texture of the wood, but lie on the surface of the substrate like paint. They are indispensible in the furniture finishing world and are particularly useful for touching up faults on a surface film. Most pigments can be mixed with either water, spirit oil, cellulose, although some blend into a medium better than others. 'Universal' pigments, which are complex metal dyes in solvents, can be added to shellacs or lacquers and used as over-tints to previously stained surfaces to obtain colour matching. A good sense of colour is needed for colour matching and it is a skill which can only be acquired through experience. In modern finishing the use of these tints in conjunction with spray equipment is an important aid to the wood finisher.

An example of how pigment is used is given in the following brief:

The brief: To re-surface a table top which requires colour matching to a given sample

The wood is solid mahogany, and a modern hazard-resistant finish is required.

After all the necessary preparation, the table has been stained as close to the required colour as possible and allowed to dry out. However, the colour does not quite match the sample.

Make up a thin mixture of the sanding sealer or base coating and mix, for example, with a little green or yellow universal pigment. It may look as thin and as transparent as tea but after drying the blending of this tint will make a perfect colour match.

You may still need to add a little pigment into one of the top surface coatings before applying the final one. This colour matching is one of the most skilled undertakings a wood finisher can perform. It takes a good eye for colour and extreme care.

Table A **Pigments in Common Use**

WHITES	Titanium White (titanium dioxide)	Very opaque with great hiding power.
	Zinc oxide	Brilliant white but must not be used under acid lacquers.
	White lead	Discolours quickly and is coarse in texture.
	Antimony White (oxide)	Used in fire retardant paints.
	Lithopone	A mixture of zinc oxide and sulphide. Is used as a substitute for white lead.
BLACKS	Gas black	An intense black solvent in water and oil.
	Vegetable black	Derived from coal tar. Good opacity.
	Lamp black	Ideal as tints for all finishes.
YELLOWS	Yellow ochre (iron oxide)	This is a dull, strong grain-obliterating colour but it is useful as an addition to shellacs and lacquers where thin match colouring washes are required. It is also useful for adding to fillers.
	Cadmium Yellow (cadmium sulphide)	This is a very bright and stable colour that is used in colour wash tinting.
	Yellow Chrome (lead chromate)	Used mainly in oil paints or oil stains as a tinting colour. Must not be used with acrylic paints(water based). Toxic.
	Raw Sienna (iron oxide)	A brownish yellow ideal for scrumbles, glazes, etc. It is quite transparent in oil mediums.
	Burnt Sienna	Pinkish red in colour with very strong obliterating power. It tends to give a chalky effect and must be used with care if added to other combinations of pigments.
	Orange Chrome	This is a very weak transparent pigment, which makes it useful for tinting other pigments. It has a rather coarse texture.
GREENS	Brunswick Green (Chrome green)	A rather weak colour used as a tinting pigment.
	Chromium oxide green	This is a very fast and chemical resistant colour.
	Phthalocyanine green	This is a metal complex pigment formulation and is ideal for any solvent in paint manufacture. It has great stability. Developed by the ICI Laboratories.
BLUES	Prussian blue (potassium chloride)	A good colour which will not fade but weak when used with alkalis.
	Ultramarine blue (sodium sulphide complex)	Poor weathering when used in paint. Not suitable under acid hardened solvent lacquers, but good colour.
	Phthalocyanine blue	A metal complex pigment formulation which is extremely stable. Developed by ICI Laboratories.
	Cobalt blue (cobalt aluminate)	A very stable blue but poor in colouring power.

REDS	Cadmium (cadmium sulphide)	A very stable colour.
	Venetian red (iron oxide)	A good stable colour.
	Red lead	This pigment is used in rust inhibitive oil paints such as metal primer. Also used with cellulose as the medium.
BROWNS	Brown umber* (iron manganese oxide)	This is a very important basic colour used in all aspects of tinting and colour washes. A good deep colour.
	Burnt umber*	A lighter colour than brown umber.
	Burnt turkey umber*	Another variation of the basic colour.
	Vandyke brown	This is a natural earth when in powder form and it can fade in this state.
	Vandyke brown crystals	A variation of the above which has been chemically treated with ammonia to produce a water soluble stain which is resistant to fading.
	Bismark brown	A strong blood red colour which is soluble in only spirit or water. It is used mainly as an addition to other pigments and for mahoganies.

* Care must be taken in using these umbers under acid lacquers as they can sometimes react with, as well as slow down drying on, the surface coating.

Stains and Dyes

Most wood finishers avoid defining the difference between stains and dyes. In reality there is none, simply a play on the words in common use. A 'dye' is a coloured substance that will 'dye' wool, cotton, wood, veneers etc, but a 'stain' will also 'stain' wool, cotton, wood, veneers etc. However, although we have been quite happy to refer to woollen dyes, it would seem strange to talk of woollen stains, but when it comes to wood it is correct to refer to both wood stains and wood dyes. It should also be noted that many manufacturers of wood colouring materials vary in the description of their own products – some state wood 'dyes' while others state wood 'stains'. What is important, however, is that dyes are the basic ingredients of the colouring matter and make up of a stain.

Universal combination dyes can be added to existing stains to improve the colour or tone of a shade. For example, if you find that a mahogany stain is too reddish, it can be toned down by adding a little green universal dye, which makes the mahogany a little warmer in colour than the original harsh red.

The principal difference between a stain and a pigment is that whereas a pigment lies on top of the surface, a stain is absorbed into the wood and becomes part of the surface area structure of that wood.

There are a number of things we look for in a stain or dye. Firstly, its solubility; secondly, the shade and strength of its colour; thirdly, light fastness, and last but not least, the chemical reactivity with its final finishing surface coating, which is particularly important when using spray or mechanical lacquer application processing.

Unlike dyes, stains are associated with the medium or liquid in which the colouring matter is held, so there are:

Colouring Wood

1. Oil stains.
2. Water stains.
3. Spirit stains.
4. Chemical stains.
5. Mixed solvent stains.
6. Non grain-raising stains (or NGR).
7. Varnish stains.
8. Preserving stains.

Oil Stains

In general, these are mixtures of analine, metal complex dyes, pitches and tars in a mineral solvent, normally white spirit or synthetic solvents, with a little drier added. These oil stains are the ones most commonly used by the wood finisher due to the advantage of having a manufactured prepared stain with a specified colour range, such as golden brown, satinwood, teak, nut brown, light mahogany, rosewood, yew, dark oak, weathered oak, repro oak, etc. These stains come with a set of wood samples so that the operator has some idea of the actual colour, bearing in mind that as most colours are shown on pine you must always try your stains on the actual substrate sample piece before proceeding. The natural colour of the wood affects the final appearance of the stain and the same stain can look completely different on different woods. Oil stains can be purchased without fear of any basic change with repeated orders. They are ready for use straight from the can and may be purchased in quantities of 125 ml up to 5 litres. To avoid problems it is advisable to stick to one manufacturer as different makers may use different solvents. Oil stains do not raise the grain of a substrate and once dry they do not obstruct or fill the grain of the wood. They are ideal for covering large areas such as wall panelling, doors, etc. Furthermore, it should be remembered that all oil stains are reversible so that they can be easily removed by using their own solvents, normally white spirit.

However, there are some problems with using oil stains and these should be understood. Some oil stains fade slightly although most are non-fade. Shellac and french polish are ideal surface finishes for these stains, but problems can occur if the stain is not allowed to dry out properly before finishing takes place. The biggest problems arise with modern surface coatings. Oil stains should not be used under cellulose or acid hardened lacquers as the contents of the stain can react with the acid used in the make-up of the lacquers; the first spray coating can bleed, burn up or strip the oil stain off the wood like a mild stripper. Furthermore, they will destroy the hazard and heat resistant quality of the lacquer. If using cellulose, a thin coating of dewaxed shellac sealer specially made for this purpose can be used to fix the stain, but it is not recommended for use under acid lacquers; that is, lacquers of the pre-catalysed and acid hardened solvent type.

Most oil stains dry slowly, but if applied at a room temperature of 65°F, they should be completely dry in 12 hours. It is a good idea not to apply more than one coating of an oil stain at any one time; a second coat will remove the first so the colour must be right with the first application. Oil stains should not be applied on oily woods such as rosewood or teak, nor should they be used under varnishes of the oil medium type as the stain will bleed.

Water stains

At one time water stains were used primarily by professionals, such as antique restorers and specialist wood finishers, but due to the increase in water-borne varnishes, paints and lacquers, these are now very important stains. They are made up of analine dyes in aqueous solutions, and can be used for general purpose staining of both hard and soft woods. Water stains can be purchased from specialist suppliers but D.I.Y. enthusiasts often find it difficult to get hold of them. Normally, they are supplied ready for use in 5 litre plastic containers but they are also available in the form of powders. The colours range from dark rosewood, old pine, yew, Jacobean, black, teak, light oak, dark oak, walnut, brown mahogany, fumed oak, medium oak, rosewood, golden oak, yellow, green, etc.

It is interesting to note that water stains are not cheaper because the medium is water but are comparable in price to oil stains. Water stains can be intermixed, so virtually any shade can be produced, and as they penetrate deep into the grain of the wood, they have a long lasting effect. One of the great features of water stains is that whilst they raise the grain of the wood, they give a transparent colour which does not obliterate the grain or muddy the colour. Furthermore, they are very stable in light and do not fade.

Water stains have one great advantage over all the others, and that is that they are ideal for use with varnishes (both acrylic and oil), cellulose, all the acid-hardened solvent lacquers, as well as oil or wax finishings, as they do not react with any of the acid lacquers and bleeding does not occur when using oil or wax as finishes.

To avoid the grain raising it is advisable to pre-treat the substrate first with water to allow the grain to rise and, when dry, to sand down, after which the water stain should be applied when the grain raising will not be as severe.

Another point to remember is that water stains should not be used on pre-fixed veneers which have an animal glue as the adhesive. However, if the glue is a water resisting type then no harm will come to the veneers. When re-veneering it is always possible to stain the veneers before gluing to the substrate, thereby avoiding problems of this nature.

One of the most popular water stains used by professionals is Van Dyke Brown crystals which are mixed in warm water with either a little detergent or ammonia to help the stain penetrate deeper into the fibres of the wood. However, when using modern finishing lacquers, such as acid catalysed lacquers, it is not advisable to use ammonia or detergents in the water stain as these two ingredients can react with the acids in the lacquer. It is also advisable to use only distilled or rain water for mixing water stains as the water supply in some areas contains chemicals such as chlorine, which again could react with the acids in the catalyst of the lacquer. When used on oily woods, such as

Colouring Wood

rosewood and teak, water stains will find difficulty in penetration so it is advisable to use hot water, not cold, in such cases.

Finally, it should be noted that the water stain must be allowed to dry out thoroughly before any attempt is made to apply a finishing coating of any kind to the substrate.

Spirit stains

Spirit stains are pigmented powders suspended in a spirit medium, but unlike other stains are somewhat difficult to apply. At one time they were not colour fast but today they are, thanks to improved manufacturing techniques.

These stains are very important and most wood finishers rely upon them for a number of different reasons. They are supplied as either:

1. Manufactured spirit stains, which are dyes in a solvent, and can be supplied either in 5 litre/1 gallon plastic containers in ready made up colours, such as golden brown, medium oak, antique oak, mahogany, oak, black, blue, bismark brown (deep red), green walnut, yellow, etc., or
2. Basic pigments, simply made up by the wood finisher with a mixing spirit, such as industrial spirit (clear), and adding a little binder, such as dewaxed shellac, and stored in a container. It is never a good idea to use fresh pigment stains, but to keep them for a few days to give the pigments time to dissolve properly. You must also make sure that the pigments you use are spirit soluble. The stains produced are heavily pigmented colours and can, if used in a strong mixture, obliterate the grain.

The standard methylated spirit (alcohol) is not really suitable due to the purple dye (blue dye in the US) contained to identify it.

Spirit stains can be applied by either spray gun, mop, brush or fad but a little skill is required. Furthermore, they must be applied along the grain, never across it. They are indispensable when it comes to colour matching as they can be made up to the exact colour and shade required by the wood finisher. One further advantage of using spirit pigment stains is that they dry very quickly, unlike oil or water stains.

Although spirit stains are more flexible in use, they should be treated with care. It is fascinating to mix your own colours and develop your own techniques, but this cannot be obtained by reading alone; hands-on experience will show in the quality of the work produced.

Pigments used in stains can be made up using either industrial spirit, which is clear, or cellulose or mixed solvent thinners. A better blending of pigment and medium is achieved if the pigment is first dissolved in the spirit to form a paste and the paste then dissolved or mixed with cellulose medium.

Sometimes it is a good idea to add a little of the spirit stain in a final spray coating to give that little extra shade tint, provided that the stains used are compatible with the lacquers being used.

An example of a colour match project is given in the following brief:

The brief: A new sideboard has been made and it needs to be coloured to match existing furniture in a room

The finish has to be a hazard-resisting finish in semi-gloss.

You have in front of you a colour match piece taken from furniture in the same room. After all the necessary preparations have taken place the first colour has been chosen (a good choice being a water stain), applied to the sideboard or a test piece, and allowed to dry. Next, you have to check the colour for matching. Upon testing against the colour match piece you find that the colour is still some way off. After experimenting with very thin washes of pigments, you settle for a match which will represent the correct combination of colour to use. This can now be made up into a fluid with very thin viscosity and sprayed on the substrate using low pressure and a fine spray pattern until matched. Once dry you can carry on applying the sanding sealer or base coat and top coatings of a pre-catalysed semi-gloss lacquer, or indeed, any other preferred lacquer.

It is possible to achieve complete colour matching by using thin washes of pigments over stains, which would be impossible by over-staining with, say, oil or water stains.

Pigments are also widely used in 'touching up' techniques, where very small drops of medium – shellac, lacquer or varnish – are mixed (like an artist would mix paints) with very small amounts of dry pigments and touched in where the colouring is required. A good plan is to have an artist's board for this purpose and small containers with the basic pigments, preferably the type that are spirit soluble.

Take, for example, the top of a bar which has a large gash in the side that has been filled but is showing the filler. Using a combination of pigments to match the surrounding wood, a colour is touched into the filler area and blended into the surrounding wood so that the damaged area is no longer noticeable. This technique requires a good eye, a steady hand and some good quality, fine pencil brushes.

Do not forget, that it is not the equipment or materials upon which you will be judged, but the final finish.

Naphtha stains

These stains are used commercially for production finishing. They give a transparent colour which is free from any cloudiness and are therefore more even in colour effect. They are also better in light fastness but are the more expensive stains in the trade to use. These stains can only be thinned down with their own compatible naphtha solvents. They do not raise the grain.

The colours available range from light oak, medium oak, dark oak, Jacobean, walnut red or brown mahogany and golden oak. Compatible naphtha tints used for colour base tints are also available in colours such as red, blue, orange, yellow, black, green and brown.

Naphtha stains are useful in that they can be used under shellac polishes and have excellent penetration in most woods, although they do have less clarity than oil stains. They dry very rapidly (bone dry within the hour) and the colours are batch controlled for quality. These stains are not recommended for spray staining or touching in, but can be applied by either mop or fad.

Colouring Wood

Non Grain-raising stains

These stains, popularly known as NGR stains, were first developed in America and are mixed solvent stains, but the dyes used are water soluble acid dyes and glycol ethers are used as the main solvents. Here again they are mainly used commercially for large scale production finishing.

They have excellent penetration powers, are fast drying, do not raise any 'fuss' in the grain, are very stable in light and can also be sprayed on to substrates. When using fillers, oil stains, for example, tend to bleed out on the filler but these stains do not.

The colour range is teak, golden oak, Jacobean oak, dark oak, light and medium oak, sapele, red mahogany, brown mahogany and walnut.

NGR stains require their own compatible thinners and are very expensive. They can also penetrate too deeply into the fibres of wood making problems later, particularly with acid catalyst lacquer surface coatings.

These stains can be applied by either mop, fad or spray gun, but it is advisable to apply a fixing base or sanding sealer by spray gun to fix the stain first.

Mixed solvent stains

At present, these are the most popular type of stains in use and are produced by most manufacturers and suppliers as own trade brand stains. They are a mixture of metal complex dyes and resins in various solvents. These stains are very light fast and, as the colour stability under acid catalyst lacquers is excellent, they can be mixed or applied under or with shellac finishes, as well as lacquer finishes, in any ratio. They also do not raise any grain fuss. The solvents are mixed – hence the name – and consist mainly of alcohol, cellulose (acetone) and naptha.

The colour range is superior to all oil and spirit stains and includes medium oak, dark oak, red and brown mahogany, rosewood, black, walnut and teak.

Due to their fast drying times, these stains are excellent for use under acid catalyst lacquers with no loss of colour. When applying the first spray coating sealer, it is better to give a thin spray-pass and, when dry, a full flood coating to avoid any slight stain bleeding.

Mixed solvent stains are expensive, but are very reliable. They can be applied by either mop, fad or spray gun.

Chemical stains

These are completely different types of stains and are unlike those mentioned previously. They consist of pure chemicals, some of which are clear, some coloured, which react upon the natural chemicals contained in the wood. The colours they produce are permanent because they become part of the wood. These stains (for want of a better word), are mainly used by professional wood finishers, antique restorers, quality joiners, furniture makers, and those in the building and contracting trades. The colours produced by chemical reactions have an advantage over other forms of staining in that they react on the wood as a whole. With oil stains, some parts of the grain absorb more colour than others leaving some areas lighter or darker than others, but this is not the case with chemical stains. The staining action is more natural and even, which is an advantage that cannot be imitated by any other stain. A typical good example of this is when dealing with figured oak; the medullary rays absorb stain at a different rate than other areas of the oak, but chemical staining is not affected this way.

Chemical stains fall into two main groups – alkalis and acids. It can be said that these stains are actually water stains as the medium is basically water, which means they tend to raise the fuss in the grain of woods. Alkalis consist of salt compounds and bases which are highly soluble in water and thus produce a caustic or corrosive solution. Acids, on the other hand, are composed of hydrogen and other elements, but, again, they are corrosive.

A glossary of chemicals used in colouring wood

It should be noted that the reaction created by a particular chemical varies according to the type of wood being treated; not all woods react to colouring in the same way.

Bichromate of potash
Copper sulphate (blue copperas)
Tannic acid
Ammonia (.880 strength)
Ferrous sulphate (green copperas)
Sulphuric acid
Permanganate of potash
Acetic acid
Pyrogallic acid
Nitric acid
Sodium carbonate (known as washing soda)
Sodium hydroxide (known as caustic soda or lye)

Chemical stains in detail

Bichromate or Dichromate of Potash is available in the form of yellow/orange/red crystals (toxic) and is one of a number of coloured chemical stains. A concentrated solution is made by steeping the crystals in warm to hot water (use rain or distilled water as the water supply in some areas may be too hard or contain chlorine), which enables the crystals to dissolve completely. The solution reacts with the tannic acid found in some woods, such as oak and chestnut, and darkens the wood within five minutes. Mahogany is also affected by this stain and darkens. It should be noted that to obtain a more even and positive staining effect, the wood can be pre-treated with a coating of tannic acid (or pyrogallic acid) and allowed to dry, and then coated with bichromate of potash. Any wood

Colouring Wood

which does not contain tannic acid, such as pine, can be treated in this way to obtain a reaction.

Copper Sulphate which is known as blue copperas, is toxic when dissolved in water, and will colour some woods a bluish dark grey. It is used principally to weaken the red in strong natural coloured red mahoganies.

Ferrous Sulphate or sulphate of iron, is known as green copperas or even green vitriol and is also toxic. This stain reacts on woods to produce a silvery grey colour, the iron salts acting on the tannic acid contained in the cells of the wood.

Sulphuric Acid used in full strength produces a light brown colour on oak, light and dark green on pine. It is also used diluted in the acid finish when french polishing to produce, with Vienna chalk, an oil free surface finish.

Permanganate of Potash consists of common violet crystals which, when used in the proportion of 2 ozs to 2 pints of soft rain water, darkens oak and ash to brown tones. It reacts quickly but it also tends to fade quickly and it is not used very much these days.

Mild Acetic Acid or White Wine Vinegar which has been poured over iron filings, nails or screws, left for eight hours and then strained off, will colour oak a weathered dark grey, even black, depending on the strength. It is quick-acting – 2–10 minutes depending on the strength and is ideal for antique distressing because it is permanent. Care must be taken to make sure that you obtain the correct colour on a test piece of the same wood, and this can be adjusted simply by adding more water or more acid.

Ammonia used in the normal trade strength of .880 or 35% pure ammonia, which is concentrated, will darken oak slightly. (As a comparison, household ammonia is about 5% strength!) Used in the 'fumed' state, it will darken oak and mahogany quickly – say 3–6 hours duration in a closed container. (In US this is known as aqueous ammonia or 26% industrial ammonia.)

Tannic Acid Oak, amongst other woods, contains tannic acid, and it is this which reacts with the ammonia in the fuming technique. Before synthetic chemicals were developed, oak bark was steeped in water as one of the only sources of tannic acid. If iron nails are used in oak, these will also react and produce a black colour around the nail head. Tannic acid in a milder form is also to be found in strong cold tea after the leaves have been steeped for an hour or so.

Nitric Acid requires very great care in use and should not be attempted by the novice. It is a colourless, highly corrosive liquid which, when diluted in a 4–1 proportion with water, produces a yellowish tone. Undiluted it produces a red/brown/yellow stain on certain woods. The most common use of nitric acid is in the pickled pine effect. By varying the strength of acid and water a weak solution gives a greyish tone to the wood. However, if used in strength a reddish/yellowish tone is obtained. Various effects can be obtained if this reaction is allowed to dry out completely and then a solution of bichromate of potash dabbed on by fad.

Another way of producing a pickled pine effect is to mix equal parts of lime and caustic soda in water (say ½ lb each in gallon of water) and apply it by brush or fad.

Pyrogallic Acid, which is an extract from gallnuts, is used to coat woods prior to fuming and will give a rich redder tone than tannic acid. It can be mixed with tannic acid to give a slightly half-way tone of a brown/red colour by the fuming process.

Other examples of chemical stains on wood are *Alkaline Solutions* such as *Caustic Soda* or *Lye* which reacts on pine and darkens it. Most will know that strong, hot or cold caustic soda also strips wood, but I am only concerned here with the staining properties of these chemicals. *Sodium Carbonate* (washing soda) will darken some woods, such as chestnut, to a yellowish brown, while *Slaked Lime* has a weathering effect on oak.

Chemical fuming using .880 ammonia (very strong concentrated ammonia)

Fuming is a darkening process caused by the fumes given off by the exposed ammonia. It was very popular during the period 1890–1930, and was mainly used to darken furniture. Today it is only used by craftsman on restoration work to achieve this darkening tone on basically English and Baltic oaks. Chestnut, rich in tannic acid, also reacts with fuming, while mahogany and walnut turn to a brown tone.

Procedure for fuming

The operation of a fumigation chamber is simple:

Construct a 'tent' of transparent polythene, or use an airtight cupboard or an old disused freezer, and place the workpiece inside the tent – for example a small table – after first removing all brasswork which otherwise would be affected. Place half a dozen small saucers filled with .880 ammonia around the work piece before sealing up the tent with sealing tape. Check this every 2–3 hours when you will notice that the oak has darkened, due to the ammonia fumes reacting with the tannic acid in the oak. The process can be speeded up by coating the wood *with* tannic acid in the proportion of 1 oz to 1 qt of water. Pyrogallic acid can be used instead of tannic acid in the proportion of 3/4 oz to 1 qt water to produce a redder tone. All woods do not contain tannic acid, but beech, for example, if first coated with tannic acid before being placed in the fumigation chamber, will darken like the oaks. Pre-coated surfaces give a more even colour than untreated woods when fumed. One great advantage in fuming is that no liquid touches the wood so the grain is not raised.

Note: Extra care must be exercised when using .880 ammonia. It is a very powerful chemical.

Colouring Wood

Varnish stains (oil solvent)

There has been a tremendous rise in the popular use of these stains by the building trades in the last few years due not only to present fashion but also a wider and better use of wood finishing technology in materials other than paint. More often than not, when you see a new property being built today, 90% of the woodwork will be finished in a varnish stain, and there are firms specialising in their production. In the UK, ICI for example, under the trade name 'Dulux', produce many highly pigmented exterior finishes. Sterling Roncraft produce the Ronseal brand, which is popular with the D.I.Y. Trade, and there are many more. However, it would be better to describe some of these stains as paint, as they are so highly pigmented they tend to hide the grain.

A varnish stain is a stain/dye suspended in a medium and, due to the special oils and resins used, it copes better with weather conditions than pigmented paint. Varnish stains do improve the look of pale and uninteresting woods; one or two coatings direct from the tin can give an attractive finish. The colour range includes the usual standard colourings, such as walnut, teak, dark oak, mahogany, medium oak and many others. The beauty of these stains is that they are very easy to apply once a good substrate has been prepared. Very little skill is required to apply them and they dry fairly quickly – between four and eight hours, depending upon the time of year they are applied. They are ideal for the building trades, cheaper type of furniture and general joinery work and are easily obtainable from most outlets selling paint. There are, however, a few disadvantages with this type of stain: it has poor penetrative qualities and should not be used on quality work; the surface can show poor colour build-up depending upon the skill of the operator during application; they do obliterate the grain to a large extent; finally, if chipped or scratched, they show up the original colour of the wood.

Varnish stains are a very popular medium to work with and with care should outlast standard painting procedures. They should never, however, be applied during wet or damp conditions or many faults will arise, and only good quality brushes should be used.

Combination stains and surface finishes that protect wood

One finish which has become prominent during recent years in the building industry represents a complete change of approach to colouring wood, and that is a pigmented varnish surface coating. These finishes are available in a range of colours including many exotic ones. When housing estates were built a few years ago, doors, windows, etc, were always painted white, but now it is more common to see woodwork on a newly-built house protected with a surface coating that shows some of the natural grain of the wood. If the surfaces on which these new finishes are to be applied are prepared in accordance with the manufacturer's instructions, then no problems will occur. A way of establishing whether a treated surface has been prepared and surface coatings applied correctly is to give it the 'cotton wool test'. Simply rub a piece of cotton wool along the surface of the wood; if the wood is smooth then no residue of cotton wool will show up upon the surface. If, however, the finish is in a very rough state, pieces of the cotton wool will become trapped by the rough fibres of the wood, which generally indicates poor workmanship.

It should be noted that some of these finishes are so heavily pigmented that they could be classed as paints. They include the normal matt, gloss and silk effects, are intended for internal and external use. The external types protect against mould growth and fungal decay.

In recent years, most of the main manufacturers have developed acrylic finishes or water-borne surface coatings for both internal and external use. The cost of these new products, however, is high in comparison to hydrocarbon solvent finishes.

Acrylic finishes

In the last few of years great progress has been made in the development of water-borne finishes, or acrylic finishes as they are popularly known. The public demand in the UK for these solvent-free, non-toxic* finishes has taken the wood finishing industry by storm but the vast cost of development and research in the field will be borne by the public, both now and in the future. These varnish stains have recently appeared on wood finishers' shelves and they are principally for interior use. The coloured stains seem muddy in appearance but on drying they clear to a transparent finish. This unique formula makes the application by brush very easy but they should only be applied to bare wood or surfaces that have been stripped to bare wood, and where there is no trace of grease, silicone deposits, etc, on the substrate.

Two or three coatings may be required to give a good finish and when dry, which takes approximately one hour, the finish is U.V. resistant (anti-fade) and low odour (almost none). One great advantage with acrylics is that the brushes used can be washed out in water as water is the 'solvent'. This applies to the clear varnishes as well. They are now available in clear, semi-gloss (satin) and gloss, and the colours available vary from one manufacturer to another. The advantage of this product is that it can be used in situations where a standard mixed solvent product would cause discomfort or distress. It is ideal for such places as kitchens, public buildings, hospitals, for people who suffer from respiratory complaints such as asthma, or any circumstances where you wish to avoid lingering odours which are often found with other types of finishes.

One well-known UK manufacturer is Rustins Ltd, who have produced a fine, clear hard-wearing acrylic finish. The surface coating is touch dry within 20 minutes and can be re-coated within 2 hours, although this depends upon the actual temperature in the area where the finishes are being applied. The finish can withstand boiling water and alcohol, making the product ideal for such items as furniture and joinery woodwork, such as floors and staircases. The finishes can also be used on cork or concrete. Important features of this product are that it will not yellow as will oil resin varnishes, and the brushes can be washed out with water.

In the US, one such manufacturer of a crystal clear 'Acrylic' varnish is McCloskey of Philadelphia. Their products are 'VOC' compliant and meet the strictest air pollution regulations of the country.

Application: The preparation of the substrate is carried out as normal, removing all grease, oil or wax; in other words, the surface to be treated must be clean, dry and smooth. When dealing with

---

* In the USA – anti VOC's (Volatile Organic Compounds)

Colouring Wood

these new water-borne finishes, the method of application differs slightly from that of oil varnish. With acrylic finishes, the milky white fluids must be applied to the substrate liberally first across the grain, and then the brush should be drawn lightly with the grain to finish off. As the solvent is water, the grain could be raised, so when the first coating has dried the surface should be smoothed down using wet and dry abrasive papers (400 grade), using a little water to act as a lubricant. Dry off and apply a second coating, followed, if required, by a third which will give you the final finish.

Acrylic varnishes contain special coalescing agents which make the particles of acrylic resin blend together as the water evaporates to form a clear transparent film.

Wax stains

These stains, which are also a new breed, combine stains, aniline dyes, lead-free pigments and wax to finish wood in a single operation. They can be used on furniture and all kinds of woodwork, including floors. A point to remember is that they are for interior use only.

The preparation of wood is essential, making sure, in the normal way, that there is no dust, raised grain or any waterproof stopper. The thick emulsion fluid is supplied in a one-pack container and can be applied to any kind of wood substrate. Those operators working with pine and requiring various effects will find this product ideal and with a large colour range. It is also available in the form of a spray wax finish, which is a mixture of resins and waxes in a solvent.

Application: Apply a liberal coating with a soft brush and work first across the grain and finish by brushing evenly with the grain. The object is to leave a good coating of emulsion upon the substrate so that the wood can absorb it. When dry, which takes about four hours depending on the room temperature, brush vigorously with a fine brush (like a shoe brush). For an extra smooth finish, lightly rub down the polished surface with 0000 steel wool and finish off with a piece of mutton cloth. Brushes can be washed out in warm water.

These wax stains are very low in odour but some contain ammonia for two reasons: firstly, to help penetration into the fibres of the wood, and secondly, to help with the darkening effect on such woods as oak, chestnut etc, which contain tannic acid. One of the fantastic uses of this type of combined wax and stain is in pine finishing, which is so popular today. The main advantage is that semi-skilled workers can accomplish fine smooth surface films, provided that the preparation has been thorough.

Stains used for modern faking or antiquing

These types of stains are for use on new leather work on furniture where a distressed effect is required. They are aqueous solvent-dispensed stains in lead-free micro-ground pigments and are available in just two shades – black and brown.

Spray stains

These stains are ideal for spray gun work and consist of analine metal complex dyes and resins in solvents. They are light fast and can be used on both hardwoods and softwoods. The colour range includes Yew, Teak, Walnut, Black, Dark Oak, Light Oak, Medium Oak, Brown and Red Mahogany.

An important point to remember when intermixing stains to obtain a desired colour for a project is to always make sufficient quantity for the job in hand; shades can vary if you under-mix or run short. Stains left over are never wasted and can be used for other jobs later on. Always store surplus quantities in glass containers and out of direct sunlight.

Stains that preserve wood

Basically, these consist of stains and a combined oil as medium plus chemicals which will beautify wood without forming a skin film. They will also prevent decay and give long lasting life to wood exposed to the elements due to their water resistant properties. Most of the man-made preservatives have a low odour, but some are high, while the new water-borne preservatives have none. The popular, manufacturered types are obtainable in either 2½ litre or 5 litre (½ or 1 gallon) containers. Following application to external woodwork, the medium evaporates leaving the pigments, together with the chemicals such as metallic naphthenates and pentachlorphenol.

These types of wood preservatives are ideal for all external woodwork, such as cladding, beams, structural timbers, etc. One advantage of using oil based solvent products is that not only do they preserve wood but they also prevent any attack by wood eating insects. They are available from a number of manufacturers in the following colours: Dark Oak, Light Oak, Walnut and Green (this type is specially made for use in the greenhouse or areas where foliage is present). When these preservatives have been used on wood, it is difficult to paint over with oil based paints afterwards unless special primers, such as aluminium primer, are used. This problem does not arise when acrylic preservatives are used.

Coal Tar wood stain and preserving oils

Creosote Oil

Although this is better known as a wood preservative, it can also be used for staining wood. Creosote oil is made from coal tar distillates and is supplied in various grades suitable for either brush, spray, dipping or pressure impregnation, the last method being the best no matter what oil preservatives are used. This is one of the oldest and most popular oil preservatives on the market today. The finish was formulated many decades ago and, in spite of more recently introduced finishes, remains popular today. Telegraph and electricity supply poles, for example, are still pressure impregnated with creosote.

Once the oil has been applied and the solvent allowed to dry out, the wood will be free from insect

Colouring Wood

attack as well as water resistant for many years. The only snag with this oil is that it has a rather powerful odour so it is only suitable as an exterior finish to bare wood. Creosote oil must not be used near growing plants or it will act as a weed killer. Pets, fish, farm animals and children must not be allowed near the oil whilst it is being applied and must be kept away until the object coated is bone dry and odourless, which may take many weeks. Once wood has been treated with creosote oil it must never be overpainted with an oil solvent oil paint or it will bleed and show through the paint. However, it will accept paint if a coating or two of aluminium primer is used, but painting over creosote is not recommended, even if several years have lapsed since the last application. The colours available are dark oak, walnut and light oak. Creosote has excellent weathering qualities, and unlike all the new water based stains and preservatives, is well tried and tested.

When using this product, great care must be taken with handling; proper clothing, goggles and gloves must be worn to prevent spillages from touching the skin or getting into the eyes as this can be both painful and dangerous.

Staining wood with any of the fluids mentioned is one of the most important processes a wood finisher has to perform. Get it right and the finish will look after itself; get it wrong and the result is misery!

Health and safety checklist

1. Most stains are best applied by rag or fad and although most of them are harmless, the dyes do colour human skin. This can be avoided by using thin plastic gloves.

2. Have good ventilation in the area when applying oil or mixed solvent stains.

3. When diluting acids, make sure the acid is added to the water, NOT the water to the acid.

4. All stains should be stored in a cool place out of direct sunlight.

5. No smoking, and no food and drink consumption whilst using stains.

6. Wear suitable clothing.

7. Wear goggles whilst using such fluids as creosote oil, nitric and other acids. Great care must be taken when using .880 ammonia (aqueous or 26% industrial ammonia (USA).)

8. Keep children, pets and fish away from the area being stained.

9. If the various materials are being used commercially, consult the UK COSHH Regulations (Control of Substances Hazardous to Health) 1988, or the American OSHA (Occupational Safety & Health Administration). These must be implemented.

CHAPTER TEN

Polishing Wood on the Lathe

Woodturning

During the past few decades woodturning has become one of the most widely practiced of all wood crafts. There are countless books on the subject and many private organisations running courses. From the early days of woodturning when chair legs were turned on pole lathes deep inside the forests to the sophisticated machines that are available today the craft has become one of the most popular occupations open to anyone who owns a lathe. Just as machines are continually changing (as we witness in trade magazines), so too are the methods of finishing the many products being produced on them.

The traditional method of finishing turned items was either by using wax or oil, paint, varnish, french polish (shellac), cellulose or simply left 'in the white'. Some of these finishes could not be applied whilst running on the lathe. Now, due to market development, many specialist firms produce finishes that are especially formulated for application whilst the lathe is *running at speed*.

Prior to any finishing the preparation of the wood must be carried out in a meticulous fashion. Abrasive papers of varying grades such as 150 down to 320 are ideal for this task. The choice of whether to use either garnet, aluminium oxide or dry lubricated silicon carbide papers depends really upon which wood is being used, as some abrasive papers can actually impare the finishing result and thus show grit markings. So care is required in the choice of abrasive material. I do not, however, recommend the use of glass paper for turning work.

One point to bear in mind is that the finish may be the *first* visual contact with an object, although it is the very *last* process applied by the maker.

Preparation

Many turned items, such as bowls and spindles, are best polished whilst on the lathe running at speed, but before any finishing can be applied the preparation must be carried out. Sharp chisels and correct use of abrasive papers – say 320 to 240 grades – are ideal to remove any traces of fine chisel marks.

Whether or not to fill the grain is sometimes a problem, but the answer is simple: if the turning is a domestic utensil for use with food, then a washable grain-filled finish is required, but if the turning – say a bowl, for example, is simply for artistic and visual use, then an open grain is better. Open grain harbours dust and grime and can be unhygienic, whilst filled grain is comparatively germ free.

The matter of choice can be further complicated by the timber which is being used. Oak and ash are much better if left unfilled, as the open grain is usually very pleasing visually, whilst mahogany and walnut, being fine close grained woods, are better filled. These are, though, largely matters of personal preference and, as always, the woodworker must decide which is preferred and also which particular finish to use.

High build Friction Polish

This is a shellac-based spirit, clear gloss finish which is better applied with the lathe running at speed. Ordinary french polish is ideal if the object you are making is for purely artistic appreciation but it is very difficult to apply by normal rubber methods on turnings. A new formulation of High Build polishes have been developed, and they are ideal for the purpose of finishing on a lathe. When the turned piece has been prepared and is ready for finishing, a good coating of the High Build Friction Polish, a milky fluid, can be applied by mop, cloth or brush.

Turn the lathe on at high speed and hold a piece of cloth charged with the fluid onto the turning. As you apply pressure the shine will appear on the surface within a very short time. When completed, turn off the lathe and allow to harden. This product is not water resistant.

Cellulose Finishes

These finishes can be brushed on with the lathe running at very low speeds or even on a stationary lathe. First of all a sanding sealer is required which is a slightly milky or semi-clear fluid that dries within ten minutes and can be sanded easily within thirty minutes. The best abrasive papers are the silicon carbide 320 grade of the self-lubricating type – 'Lubrisil', manufactured by English Abrasives, for example. This is far superior to many other kinds of abrasive paper for woodturning work. On woods like elm, lime and mahogany, two or more coatings are advisable to obtain a good build-up of a base surface coating of sanding sealer.

Cellulose can also be applied either by spray gun or aerosol cans, or alternatively, by dipping. It all depends on the quantity of turnings in hand. If you have thousands of axe handles to finish, then a dipping tank is ideal, but if you have only a couple of bowls to finish then one of the other methods of application would be better. In commercial turning, time has to be the paymaster, but when turning simply for a hobby this is not a factor of great importance.

When the sanding sealer has dried, one coating of cellulose clear lacquer, either gloss, semi-gloss or matt, is required. This is best sprayed on if possible, preferably off the lathe, and then allowed to dry. When bone hard in approximately six hours, depending on the temperature in the workshop, the turning can be placed back on the lathe and run at slow speed to denib any faults using worn 320 Lubrisil abrasive papers. If a semi-gloss finish is required, use a little 0000 steel wool gently on the surface until the desired effect has been achieved. If a full gloss is preferred, then after denibbing use a french polish type rubber but with a little pullover fluid, and apply this to the turning at slow speed and a full gloss finish will result. This can be further improved after two hours by burnishing the turning at a high speed using a burnishing cream. This will give a fantastic gloss finish.

After the evaporation of the solvent, the thin film tends to shrink and this must be allowed for before burnishing. The resultant finish will be hard, easily cleaned and have hazard resistant qualities. Acid finishes (i.e. the two-pack lacquers) are not recommended for use on turnings.

Wax finishes

Beeswax is useless on turning because it is too soft and collects dust and grime. The much harder carnauba wax, if held against the turning with the lathe running at high speed, will give an excellent finish to spindle work. The technique is to hold the wax hard onto the turning, allowing the heat (friction) to slightly melt the wax. Then apply a 320 grade Lubrisil abrasive paper to the spindle when turning at speed which actually fills the grain at the same time. It is the combination of the carnauba wax and the abrasive paper forcing it into the grain of the wood that gives it the fantastic smooth quality antique finish. If you still want to seal this finish, just give it a thin coating of clear friction polish.

Varnish finishes

Varnishes have their place when used as finishes for turnings. They can be the standard oil solvent types, either clear or combined with a stain, or can be one of the new breed of acrylic varnishes now in matt, semi-gloss, gloss or satin finishes.

These water-borne varnishes can be applied to bare wood, such as turning, whilst off the lathe, followed by one or two more coatings, and, when dry can be put back on the lathe at a slow speed to denib, using either steel wool or 320 grade Lubrisil abrasive papers. These finishes give a very durable finish to precious woods and are odour-free. Due to their formulation they also reduce the penetration of ultra-violet light and prevent darkening on such woods as pine, which the oil solvent types are known to do. The oil solvent types also take many hours to dry out and harden, and for this reason I do not recommend them for turnings.

One such UK brand is manufactured by Cuprinol under their 'Enhance' trade mark. In the US one of the brands is McCloskey's clear acrylic varnish.

Oil Finishes

Oil finishes, such as linseed oil (raw) and teak oil, were once very popular, but they dry out and become engrimed and require constant re-oiling to keep their patina, and whilst they do enhance woods this effect is short-lived. A product that has been on the market for many years, called Danish Oil has become established as a first class finish for wood turnings.

Polishing Wood on the Lathe

Danish Oil (Rustins)

Danish Oil is a very different kind of oil in that it does not leave a thick skin like boiled oil or pick up dust and grime like linseed oil. On a woodturning the oil can be brushed liberally all over the surface and will be dry in approximately four hours. When completely dry the surface can be buffed or burnished using a slow speed on the lathe to produce a full lustre to the wood, which is also a durable and water resistant finish. This oil is ideal for all turnings and is manufactured by Rustins in the UK. There are other firms on the market producing similar oils but for this type of work I, personally, have always used this particular product. It is also used extensively by many wood turners and teachers of the craft.

In the US Benjamin Moore manufacture their Scandinavian oil finish which produces a mellow sheen.

Fig 27   Rustin's Danish oil – an ideal finish for turned woodwork

Stains used on wood turnings

Aesthetically, there are purists who object to stains being used on woods of any kind. However, stains do enhance the grain of the wood and using exotic colours such as greens, blues, reds etc, can achieve wonderful effects on wood. The danger is not knowing what stain to use resulting in the possibility of a great deal of time and material being wasted.

If using shellac finishes, then mixed solvent or oil stains can be used provided that the stain has completely dried out. Then a thin coating of a shellac de-waxed sanding sealer is applied to give a good base coat to the wood. When this has dried, the top surface coating, such as shellac varnish, wax or High Build Friction Polish can be applied.

The danger here, however, is that any stain with a medium of oil such as the standard oil stain or mixed solvent oil stains can bleed through the top surface coating, such as with oil finishes. The correct stain to use, therefore, is a water based one. This slightly raises the grain, so keep the application as dry as possible and do not swab the stain onto the turning, but simply allow the minimum amount of water onto the wood. This can be quickly dried out by running the lathe at high speed and holding a cloth or fad against the wood. The friction causes the wood to dry out rapidly, causing the minimum amount of grain fuss to rise.

Chemical stains can also be applied to turnings but these must be applied whilst the lathe is at stop (see chapter on stains) to allow the chemical reaction to take effect. When dry, final finishing can then take place.

Safety in the woodturning workshop

Although the regulations apply to employers, the self-employed and any visitor or full or part-time employee using such items as a woodturning lathe are at risk and are under the responsibility of the owner of the workshop. However, most woodturners are hobbyists and the regulations may not concern them, but good practices while working with a lathe is paramount to good workmanship.

The important health risks when turning some exotic woods are that they can cause acute problems to sufferers of asthma or similar respiratory conditions. By using the correct equipment, turners can pursue their hobby or trade without the same hazards.

First of all dust extraction is very important. One efficient system is the 'AEG Compact' which consists of two motors and filters built into a drum and designed for heavy duty situations where total mobility is required. It may also be used in conjunction with a plastic bag filler which collects all dust or shavings. This kind of equipment keeps the actual working area relatively dust free, but while this takes care of removing unwanted dust particles, goggles must be worn, or better still face screens which enable people who wear glasses to see in safety when turning wood. These face screens are superior to goggles, giving better visibility and safety.

Powered Respirators

For operators who turn wood for a living, or for anyone else who works with wood, there are powered respirators available which completely cover the worker's head in a visor hood, and which uses a battery controlled power pack to supply clean air through charcoal filters over the face without misting the visor. This system is especially suitable for spectacle wearers, and complies with safety regulations.

A Check list for health and safety

1. Never work with long trailing pieces of cloth or muslin. They could become entangled with your fingers or the lathe, or both, whilst the lathe is at speed.

2. When using steel wool for dulling or denibbing, always return the swabs to a metal tin containing water. Friction causes steel wool to ignite and glow slowly red hot which could cause a fire hazard. These swabs can be dried out at a later time for re-use. (I, personally, have seen steel wool glowing red hot which was started by a single spark from an electric motor.) Great care is required when using steel wool on lathe work.

3. Have plenty of ventilation available whilst using mixed solvent stains or cellulose sealers or other finishing coatings, oils, french polish, even waxes, or you could end up with painful, sick headaches.

4. If sanding large production pieces, wear a dust mask and head covering. Dust can be a health hazard, and the dust here is not always from the wood but from the surface films. Preferably have dust extraction near the lathe.

5. Do not smoke or consume food or drink whilst working on or near the lathe or whilst using stains and finishes. Do not forget that the cellulose finishes, methylated spirits and thinners have low flash points (fire hazards).

6. Do not have open flame heaters, such as butane gas heaters, near to where these stains are being applied.

7. Wear protective face shields, goggles, etc, when woodturning. Also wear a woodworkers smock, overalls or apron. Avoid loose sleeves, ties or other garments.

CHAPTER ELEVEN

Shellac and French Polishing

To wood finishers the world over, shellac means traditional wood finishing. Many craftsmen and women, furniture restorers, contract polishers, builders, joiners and also many industrial processes including the printing trades, use shellac in some form or other, as do others far too numerous to mention here.

'Gum lac' (not shellac) was used by such early civilizations as the Chinese, who dyed silk and leather with the rich red dye which is still today a waste product from the process of washing the basic raw material. The Egyptians used gum lac to colour furniture as did the later Roman and British Empires who also used this lac as a popular natural adhesive as well as a colouring agent. The British Army uniforms (redcoats) were once coloured with the dye of this product. Gum lac was one of the many products exported to Britain and generally distributed by the East India Company which flourished from 1700. And later, the industrial revolution of the early 19th century which swept the western world ensured an increasing need for the product mainly as a colouring dye and an adhesive.

But what has shellac to do with gum lac? To understand what this wonderful material is and to know its properties, and why it has become so important to our wood surface coating manufacturers, a little further history will be required here. In 1855, two men known as the 'Angelo Brothers' opened the first factory in Cossipore, near Calcutta, to produce and export in a more commercial way the lac dye which was in demand throughout Britain and Europe. In 1856 the process of using the raw material supplied by native cottage industries to the new factory at Cossipore was developed, and the product was used to produce both dyes and the gum adhesive. However, due to the discovery of a much cheaper way of producing colouring matter for the cotton trade by 'Perkins', who discovered aniline dyeing, in Europe the use of lac dye declined and the Angelo factory turned its efforts in 1872 to producing a new type of product called shellac by using a new solvent method in its production. 'Shellac' is a refined form of lac with the word originally deriving from 'shell-lac' – the thin flakes in which form it is commonly marketed throughout the world.

Development and use of the new product co-incided with the vast expansion in Europe of an entirely new way of finishing wood and furniture – 'French Polishing'. To a world where wax, oil and varnish were virtually the only finishes, it was equivalent to our present day technology producing the new acrylic finishes in the field of paint manufacturing.

French polishing had arrived and the cabinet trades soon turned to this new way of finishing their products. Old established companies such as 'Broadwoods', the famous English piano and harpsichord manufacturers, turned entirely to french polishing all their products from 1815 onwards. Wax polishing as a commercial finish was doomed to obscurity.

The Angelo brothers were in the right place at the right time to develop the production and export of shellac and India is one of the world's leading exporters of the product which now serves many industries, quite apart from the paint and varnish makers. It is used in such products as hair lacquers, paper finishes, adhesives, printers' inks, leather dressing, floor polishes, pharmaceuticals, confectionery, dyestuffs, plasters, etch primers, metallizing paper, sealing wax, electrical components, rubber compounds, gasket cements – the list is endless.

Whilst India is a main exporter of shellac, it is also 'cultivated' in Burma, Thailand, southern China and Sri Lanka. The whole cultivation is dependent upon the scale insect, *Laccifer lacca*. In the past, this insect has had a number of names such as *Coccus lacca* and *Tachardia lacca*, but entomologists have given it the family group name of *Laccifer lacca*. The male of the species is bright red in colour, but it is the female that is the producer of the lac (resin). They are about 3mm/7/8" in length when fully grown and attach themselves to the tender twigs of the tree. These insects are actually parasites, but infestation is deliberately encouraged by tying bundles of infested twigs, called 'broodlac', to uninfested trees where new and tender shoots are plentiful. Suitable host trees include palas, ber and kusum, and in Thailand, the rain tree.

The larvae settle on the twigs in large numbers of about 100 to 150 per $25mm^2$/sq.inch, and they feed on the sap juice of the twigs of the tree which quickly become covered with layers of lac secreted from the glands under their skin. The secreted lac resin builds up on the twigs and thin branches and gradually hardens and thickens as more and more lac is secreted. After about six to eight weeks, the larvae mature, and there is a rapid increase in lac secretion and at this time the female insect becomes purplish-red in colour. It is this colouring plus the encrusted scales of the insects which gives the lac its characteristic sealing wax appearance.

The encrusted twigs or thin branches are then cut and gathered into 'stick lac' bundles. The resin at this raw stage is composed of 70% lac resin, the rest made up of wax gluton and colouring matter. Cultivation (or lac farming) is carried on by local people and based as a village industry. The native cultivator either owns his own trees or rents them. The crop or raw material of encrusted sticks is known as 'stick lac'.

The lac at this stage is full of impurities and is refined in the first instance either by primitive methods in the villages or by more modern mechanical methods in the factory, by crushing, sieving, winnowing and washing out the dye to produce a semi-clean refined product called 'Seedlac'. This can be further refined by stretching it into sheets which when broken up into tiny flakes, becomes 'shellac'. Further purification can be carried out by various modern methods or by hand, but the result is the same in the end – shellac.

The centre of this very important industry is Calcutta, and the many variations in the colour of shellac depends, like tea, upon where it is cultivated. These colours can be golden yellow or yellow to reddish, while the Chinese shellac is predominently dark red. Shellac can also be processed by bleaching methods to produce a pale colourless shellac, and the use of various solvents can produce a colour range varying from ruby red (Garnet) to super blonde. The blonde varieties are not only de-waxed but made colourless by the process of activated carbon techniques. Thus, A.C. Garnet could mean (Angelo Cossipore Garnet) or Activated Carbon Garnet – either way it means that they are de-waxed. One important use for de-waxed shellac is in sanding sealers for use under french polishes, varnishes and cellulose lacquers. Shellac contains a great deal of lac wax and this itself has many uses in the process of polish making – for example, shoe polish.

Shellac and its use by the wood finisher

Shellac flakes, when mixed and dissolved in either methylated spirit, finishing spirit or methylated clear finishing spirit produces a product which, when applied to wood, dries out within an hour or so to a hard surface coating. This mixing process should be carried out in glass or plastic containers. Shellac, thus mixed, can also be stored in metal containers, but certain chemical inhibitors are required to be added to prevent corrosion and discolouration of the fluid when in contact with the metal.

There is a degree of confusion about the difference between shellac and french polish. The latter is made up with shellac flakes dissolved in methylated or industrial clear spirits, with additives such as gum arabic or gum copal, and sometimes a little cellulose to make it slightly water resistant and also to flow better. In the past, wood finishers made their own french polish – and indeed some still do – usually by simply adding shellac flakes to methylated spirits, and adding their own flow ingredients. The contents require a few hours to dissolve. Nowadays, however, the product is so readily obtainable from most paint and hardware stores and trade houses in various forms, that the effort is hardly worthwhile. However, it is somewhat cheaper in price.

There are multiple choices of wood finishes using shellac as the main ingredient, and to the wood finisher the term 'cut' or shellac is important because it conveys the amount of shellac flake which has been dissolved in a gallon or 5 litres of spirit. The normal cut of shellac is 3 lbs which gives a fairly thin mixture of french polish, while an 8 lb cut would mean a very heavy or thick mixture. It all depends upon the job and the craftsman's choice of french polish in conjunction with the project in hand.

Button Polish is an orange full-bodied polish ideal for most warm-coloured wood such as walnut, mahogany etc, while *Pale Polish or Extra Pale Polish*, which are near-clear fluids, are ideal for woods that are to be polished to allow the full beauty of the natural grain and colour of the wood to show through, without any alteration by the polish. Elm for instance should be polished with pale polish and not button polish. The ever popular *White French Polish*, which looks like milk, is ideal for such woods as pine as it inhibits discolouration, or for any wood that must have a natural finish, such as lime, etc. White french polish is made from bleached shellac.

Garnet Polish is a dark-coloured polish and is ideal for all dark woods such as rosewood, mahogany, ebony, etc. Garnet Polish is often referred to as A.C. Garnet – this is a polish de-waxed during the manufacturing process – the shellac is de-waxed but also decolourized by the process of Activated Carbonisation. Garnet polishes are ruby red.

Coloured French Polish, such as black (for piano finishes), and for the modern market, blues, reds, greens, etc, can be made to order by various manufacturers, or tints can be added to pale polishes depending upon using compatible finishes.

Shellac sealers have become popular during the past few years: these are generally bleached and de-waxed shellac products and make an ideal bodying surface undercoating.

Shellac and French Polishing

Knotting is a de-waxed shellac product used for covering knots to prevent the sap causing damage to the surface coating which is sometimes called knot-weeping. It is used mainly under oil surface coatings (paint).

The advantages of shellac for the wood finisher are many. It can make an excellent sealer for the surface of wood (in the white), and can be built up and flatted down between coats many times in the space of a few hours ready to take other finishes such as varnish, although it must be pointed out here that shellac must only be used as a sealer under interior varnish and never exterior varnish use, regardless of quality. Shellac with gums added is also used as a final finish in the form of french polish to quality furniture with either a full gloss or a matt finish. It is used as a spirit varnish in its own right, and can be mixed with a proportion of plaster of Paris, linseed oil and methylated spirits to produce a quick-drying filler.

The touching-up qualities are excellent. On french polished surfaces, by using a little shellac plus methylated spirits and pigments such as brown umber etc, the mixtures can be used for high-lighting on surface coating, around knots, etc. The surface of a french polished finish is easily repaired, and faults such as rubs, burn marks, scratches, white rings and so on, can be easily removed, whereas a similar fault on a modern surface is much more difficult to eradicate.

A french polished finish can be applied either by mop, brush, rubber or spray gun, yet the durability and strength due to its flexibility is long-lasting. Shellac, despite its many admirable properties does, like most other finishes, have faults, and these the wood finisher must understand. The finish is not water resistant, nor will it stand heat in any form – either dry or by moisture, and should not, therefore, be used for such items as bar tops, kitchen furniture, bathroom fittings, etc, or it will turn white and streaky. It must not be applied at temperatures under 60°F or in damp condition or blooming will take place, nor should it be applied as an exterior surface or sealer under, say, varnish, or the finish will quickly deteriorate. Added to this, it also has a limited shelf life. Special exterior french polishes, however, have been developed which are suitable under these circumstances.

Shellac, in recent years, has undergone a fantastic price revision due to a number of factors. Recent climatic conditions have not been favourable for the insects – for example, areas like Madhya Pradesh in central India have suffered droughts, and there have also been problems of cultivation because of the lack of incentives offered to the lac farmers. There is also an increasing demand for the product by newer industries other than the more traditional paint and varnish users, and all these factors have combined to cause a dramatic jump in the price of shellac-based materials, so that they are now more comparable with cellulose-based products.

Modern Shellac Finishes

In addition to the french polishes already described, there are, in many of the specialized trade house catalogues, french polishes that can be applied to wood by spray gun. These are the new breed of finishes required in commercial finishing. They save on time yet produce a good finish, but once again, not forgetting that good preparation is the key to success.

The types available for spray application are: Transparent, Pure Button, Button Polish, (these have extra shellac or body), Spray Garnet, and a spray button polish for exterior use. For metal there are shellac metal lacquers in the following colours: Cold Blue, Cold Silver, Green Bronze, Black and Gold, and these are generally available from specialist finishing suppliers.

For use in conjunction with the above finishes, there are shellac sealers which are mixtures of de-waxed shellac and/or synthetic resins, film formers and metallic soaps in solvents. These when dry provide a good base surface film which is easily sanded or flatted, and they are also fairly fast in drying. The sealers are available as either Natural, Clear Orange, Brown, Dark, White (clear) and they can be used under many surface films such as varnishes, cellulose, pre-catalysed lacquers and french polish.

It must be noted that standard french polishes contain lac wax within the shellac (approximately 4–5%) and on no account must they be used under any of the chemical lacquers, particularly the acid catalysed lacquers, because they are not compatible due to the wax content, and can therefore cause surface faults if so used. If a french polish, for example, is used under a pre-catalysed finish it will damage the anti-resistant quality of this lacquer.

French Polishing – the Traditional Method

No two polishers carry out an identical procedure of this craft and a good deal of skill is required to apply this finish to produce a quality result. Generally, it is a finish which is nowadays restricted to expensive items of furniture and woodwork, and for refinishing antiques, etc, requiring a mirror gloss or antique finish.

The materials required to french polish are:

1. French polish. (Alternatively use button polish, pale polish, white polish or any coloured french polish.)

2. A piece of wadding and a square piece of well-washed cotton or linen (to make the rubber).

3. Linseed oil (raw), mineral oil or poppy oil, and methylated spirits.

4. Finishing spirit (clear).

5. Stains, grain fillers, stoppers, abrasive papers (320/240 grade), sanding block.

6. Cabinet scraper.

7. An airtight container in which to store the rubbers to keep them moist.

The whole method of french polishing is based on the rubber. My way of making one is to take a square of well-washed cotton or linen sheeting which can vary in size between 15 × 15 cms (6" × 6") to 25 × 25 cms (10" × 10") according to the job for which it is required. Into this, place a piece of wadding around which the cotton piece is wrapped tightly into a pear shape. Some polishers fold the surplus sheeting on top, whilst others twist it tightly. Each produces a rubber which best suits their method of working, which 'feels' right, and achieves the results they want.

The purpose of the wadding pad is simply to act as a reservoir for the french polish and nothing

Shellac and French Polishing

else. It is the cotton sheet that is of the most importance in the making of a good rubber. The rubber is charged with polish by undoing the pad, pouring the correct quantity of polish (this will come with experience) on to the pad, then re-wrapping it into the pear shape. On no account must the whole pad be immersed in polish for charging. The size of the rubber is dependent upon the job in hand: if the job is a large door then your rubber would need to be large, or if the job is a small antique chest, then the rubber would be small. Once a rubber is made it will last for many months, even years. Now and again they can be washed in spirit to keep clean, but most important, they must be kept after use in an air-tight container or they will very quickly harden and are then useless.

When the rubber has been charged with polish, it must not ooze with polish or be too dry. A slight pressure on the top of the rubber will force the polish to flow through the cotton cloth and allow a thin film to spread over the substrate. This is basically the action of the rubber. When re-charging the rubber, it is always a good plan to change the position of the wadding as this prevents the sole of the rubber from becoming skinned over, and so becoming hard and useless.

The method of french polishing

French polishing is divided into three main phases:
a) Skinning in;
b) Bodying in;
c) Spiriting out.

The most important factors when polishing are that there must be a dry temperature in the region of 65–70°F, a dust-free atmosphere and no draughts or damp moisture in the atmosphere.

1. The first step in french polishing is to prepare the substrate by removing all wood surface faults, filling any holes with stoppers, and, by using a cabinet scraper, abrasive sheets, etc, produce a fine, smooth, dust-free surface. Then apply a light wash over with warm water which slightly raises the grain of the wood. When dry, sand down using 240 grade garnet abrasive papers, and dust off.

2. Apply a coating of stain as required – water stain is better in this instance, and allow to dry (do not sand at this stage).

3. Apply a grain filler or a colour as near as possible to the stain. Allow to dry and flat down using 240 grade abrasive papers (Garnet).

4. Apply a second coat of stain. This could be the original water stain, but an oil stain could be used here instead. Allow to dry.

a) Skinning In

Charge the rubber with polish and as quickly as possible apply a layer of neat polish in the direction of the grain and leave to harden for approximately one hour. When dry slightly flat down with used fine abrasive garnet papers simply to remove any nibs in the polish. It may be necessary to apply a further coating of polish at this stage. Allow to dry.

b) Bodying In

This is the main part of the operation and consists of applying various coatings of french polish by the application of the rubber covering the whole of the surface, replenishing the rubber as required, and polishing in circular and figure-of-eight motions. It is here that the use of a little lubricating oil touched upon the heel of the rubber is required to help prevent the rubber from sticking to the surface, but only a *very little* oil should be used. The process of bodying in may take many applications and the surface should be worked on for no longer than fifteen minutes at any one time – this enables the polish to harden. Over-polishing before each layer of polish has had a chance to harden can lead to a serious surface fault. A little and often is the rule! It must also be stated here that after each session of polishing, the surface, when dry, must be flatted down smooth so as to remove any nibs, etc. This can be achieved by using 240 or 320 grade abrasive papers (garnet or Lubrisil). The object so far is not to produce a gloss finish, but a good, sound, firm, flat surface with the polish evenly applied.

c) Spiriting Out

This is the stage to remove all traces of the oil that was used as the lubricant in the bodying-in process. Methylated spirits (alcohol in the US) can be used but it is far better to use the clear 'methylated finish' or 'finishing spirit' (this is a clear solvent with added resins) which is specially prepared for the job, and which allows the spirit to flow more easily than methylated spirits.

The method is to take the rubber, either the one you used to polish the substrate, or a new one, and apply a little spirit to the wadding and cover with the cotton cloth. Gently squeeze out the surplus so that it feels cold and damp on the back of your hand, and then, in straight movements with the grain, apply the rubber back and forth along the surface. This must be done in a gentle way and the oil will disappear leaving you a shine which is free from oil smears. I must point out that this method takes a little mastering and it is only with practice that the quality of work is achieved.

There is also another method of removing surplus oil from a french polished film which greatly improves the final surface of the shellac film that has been applied throughout the bodying-in stage. It is described as the antique patina finish, although it is sometimes called the 'acid finish' or 'piano finish'. This is a professional method of producing a perfect, flat, full gloss mirror surface film of superior quality which can only be achieved by this method of hand french polishing, but it is not a method for unskilled hands to attempt.

The Beezer Method

The surface must be first flatted by the use of a 'beezer'. This can be made easily with a length of thick felt cut to a width of approximately 50–75mm (2"–3") or even wider, which is then rolled up tightly to a diameter of 75–100mm (3"–4") and fastened by a wire or nylon cord in the middle of the roll to hold it together. The whole thing is soaked in either raw linseed oil or mineral oil for a few

Shellac and French Polishing

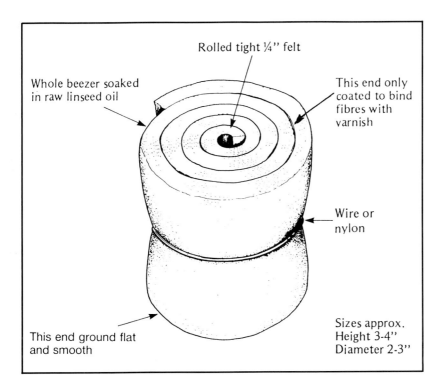

Fig 28 The 'Beezer' – used in flatting a french polished surface. This should be kept in a sealed container when not in use.

hours and allowed to dry. The top end of the beezer is sealed by pouring on an oil resin varnish and allowing it to harden. The other end of the beezer is then ground perfectly flat by grinding it on a rough stone or sandstone, using a lubricant, until the result is a firm, smooth, flat surface. When this has been made it will last for years.

The polish film is smeared with raw linseed oil and a dusting of fine pumice powder is sprinkled onto the surface. The beezer is then used as a grinder which flats the surface in conjunction with the mixture of oil and pumice. The method is to press hard on to the beezer in a rotating action, making sure that there is a good supply of oil on the surface. What happens is that the flat edge of the beezer, with the compound of abrasive oil and pumice, grinds flat the polish film which is not possible by any other means. On badly flatted surfaces the beezer flattens out what can be described as a ploughed field effect, which often happens when over-polishing with the rubber. When finished the residue is wiped off the surface using white spirit and allowed to dry, making sure that there is no oil or pumice left upon the surface.

The pumice can be kept either in a muslin bag called a pounce bag, which is then 'pounced' or dabbed upon the surface (this can be done before the oil is applied if preferred) or it is easier to use a small container with numerous small holes like a pepper pot through which to dust it on. (I find this way much better than a pounce bag). If the surface now requires further applications of polish or

Fig 29 French polishing an antique chest of drawers

bodying in, this can be done, but the polish should now be reduced by the addition of methylated spirits in a proportion of 50% methylated spirits and 50% polish, and the surface must be left in as perfect condition as possible. It is most important that no attempt should be made to 'bring up' the gloss at this stage.

It is now very important that the whole surface hardens off and this can take at least 24 hours at 65–70°F.

The Acid Finish

The oil now has to be removed from the surface polish in order to produce the full quality gloss. (The oil by now has to a degree become buried in the film of shellac and has to be removed). This is done by the use of a new rubber made in the same way as the original one used, only, instead of using french polish, the wadding is soaked in mild diluted sulphuric acid, a dilution of, say, seven parts water to one part acid. The solution is applied lightly all over the surface and immediately dusted with *vienna chalk*, which is precipitated chalk containing magnesia, and again a pepper pot container is ideal here. Using straight movements with the rubber in the direction of the grain of the substrate, continue to remove all traces of oil deposits from the surface of the polish. The oil is removed by the chemical reaction of the acid with the vienna chalk which also acts as a slight abrasive. It takes some while to achieve the result by using this method, but eventually all traces of oil are removed and what you have is, a) a positively oil free surface film, and b) a brilliant full gloss mirror finish. The vienna chalk can also be used just by itself if required to slightly burnish any french polished surface film without fear of damaging it in any way.

Shellac and French Polishing

What I must point out here is that if there is any fault in the finish caused in the bodying-in stage, or any other way, this fault will show up on the finished gloss, and I cannot emphasize too strongly the importance of preparation in all the procedures leading up to this method of finishing off.

A semi-gloss french polish finish

This is what can be described an an antique finish, which is achieved by first french polishing up to the bodying-in stage, and then leaving to dry hard for 24 hours. Then, using fine steel wool (0000) with a good quality wax furniture polish, rub well into the hard film of polish in the direction of the grain, and then finish off with a clean piece of mutton cloth. This produces a very good quality semi-gloss finish which looks mature, but it must be done with care.

Hand french polishing to produce a first class finish is a time consuming occupation. At the turn of the century, time was not as important due to cheap labour and a different mode of living. Today all that has changed and skilled labour costs money. In the first class furniture reproduction making industry, french polishing is still of vital importance to the quality when finishing such items as furniture and quality joinery. By using modern methods and equipment a comparable result can be

Fig 30   Hand finished french polished surface. This upright piano is an example of an antique that has been refinished with french polish applied by hand rubber in the traditional manner, using traditional methods.

obtained to that produced by traditional hand methods, yet in no way interfering with the surface quality.

Method of Modern French Polishing Techniques

I have no need now to impress on the reader the value of surface preparation. This must be achieved and the colouring stains applied as in the normal procedure for a traditional hand finish. It is at this point that we part company with tradition.

1. Apply to the surface substrate a spray finish of de-waxed shellac sealer, either the clear or brown, applied by normal spray gun methods, and allow to dry hard. (Note: Spray gun finishing is fully covered in Chapter 15.)
2. Sand down to de-nib, using fine 320 grade Lubrisil abrasive papers and dust off.
3. Apply further coatings and de-nib until the required surface body film is achieved. It must be stated here that the special spray shellac polishes only should be used for this method as they are manufactured for this purpose. When the required body of shellac sealer is satisfactory and the surface film is perfectly flat and free from any surface faults, allow to dry and harden for 10–12 hours, or overnight.
4. Apply the surface coating of spray-on french polish by spray gun. This must be carried out in a dust free controlled temperature of approximately 65–70°F, with no draughts of any kind, or blooming can occur. Apply no more than three top coats of french polish and allow to harden off. Each coating must be flatted before respraying.

From start to finish, the job can be carried out to a good standard of workmanship in approximately two hours, excluding drying times. It is here that a good quality spray gun operator knowing his or her polishing techniques can produce very easily a good quality french polished finish for commercial acceptance.

Yet another commercial method is to apply the sealer coatings with a spray gun, and *then* carry on using the traditional method leading up to the acid finish and so on. There are many combinations to improve on time and labour costs. The important point to remember here is that it is the resultant finish that counts.

Brush on french polish

Many firms are producing a 'French Polish' which can be applied by a brush. These are a new concept of finishes for interior woods with the name of either Brush-on French Polish or Brushing French Polish. I sometimes wonder if the name of these finishes are somewhat misleading, however the name implies that they can be coated by brush.

The properties of these new finishes are of a pale transparent type 'varnish' which produce a clear film 'similar' to a traditionally applied french polish. Now, I am not going to insult the reader by saying that these finishes are for those who cannot french polish, but they are for situations where any user can produce a reasonable finish easier than by normal traditional methods.

The method of application is completely different from traditional french polishing, and this product is mainly for new wood surfaces, such as in kitchen and bathrooms, oak beams, wood block floors, stripped pine furniture and so on. The first coating dries within 15–20 minutes and this can be followed by at least two other coatings. Flatting is carried out after each dried coating stage as normal procedure. When the finish is completely dry the surface can be burnished with fine steel wool and a wax polish to produce a fine smooth satin finish.

These finishes are resistant against some of the normal hazards on household furniture, such as white rings, etc. They are produced by various manufacturers, one such in the UK is John Myland who produce 'Brushing French Polish'. If one accepts the limitations of their use they can be very adaptable where speed and unskilled hands require a reasonable finish and therefore I welcome this newcomer in particular for the D.I.Y. Trade.

Fig 31 One of the new brush-on 'french' polishes

Health and Safety whilst french polishing

1. Do not smoke or have open flame heating equipment.
2. Do not drink or consume food in the workshop.
3. Have good ventilation through the working area.
4. When sanding or flatting, wear face masks.
5. Use hand barrier creams before working.
6. Keep items such as acids in special separate storage containers and in a lockable cupboard.
7. Keep all bottles etc properly labelled.
8. When using acids, wear protective gloves.
9. Keep animals and small children well clear of the workshop or working area.
10. Keep all bulk stock in metal cabinets.

Figs 32–33   Hand polished period chairs refinished by the author

CHAPTER TWELVE

Distressing and Other Exotic and Simulated Finishes

Distressing is a 'nice' alternative word for faking, although distressing is a much better word and has a different ring to it. Personally, I do not like the word fake, as it implies that it is some shady handy-work intended simply to cheat or deceive.

To make furniture or joinery look older than it actually is must be carried out properly and in my view becomes, if well done, an acceptable art form which cleverly deceives the eye.

For example, when a beam has to be replaced within a 16th century cottage due to extreme woodworm or decay, it is necessary to either obtain a replacement of suitable age – something which is not always easy, or to install a new beam which in its modern manufactured state would look completely out of place. Here is where the craft of 'faking' or distressing that beam to look like the others in the room must be carried out with skill, knowledge and minute attention to detail, and any new repair must blend in with the surrounding woods. When the job is complete, it would be compatible with the rest of the room, and no-one would be any the wiser. That is what good distressing means!

There are today, in any large D.I.Y. store supplying building materials 'take-away beams' made of plastic, which simply glue or screw on to existing woods, and very good some of these can be, but they give themselves away due to the evenness of their colours and can look cheap and nasty. Even these, however, can be improved by putting a little extra work into them, such as varying the colour, using a little matt varnish here and there, and a little white pigment glaze.

The practice of thinking that all distressed work must be dark is also a common mistake. New pine match-boarding can be made to look extremely old by giving a white pigmented thin glaze wash to the pine, followed by areas of green and brown umber to simulate mould or fading, and you can even include fake woodworm holes.

Years ago, I carried out a job distressing a brand new kitchen for a client, but when he later came to sell the house, the buyer thought that the kitchen was so old that it needed to be replaced with new!

One thing that you need in order to do this type of work successfully are original examples of old woods, and these are most often obtained from old buildings or pieces from furniture; alternatively, good close-up photographs of old weathered woods or panelling are also an added help.

In furniture restoration, as well as structural work, a new piece of wood sometimes has to replace a worm-eaten one, and then has to be distressed to hide the newness. When choosing a compatible replacement piece of wood, match it with the running grain as nearly as possible, and then carry out the distressing on the wood, followed by the distressing to the finish itself. Within a very short time the repair will become compatible with the rest of the item of furniture, and blend in unnoticably with the whole.

A close study of old furniture or old building construction, and visits to local museums help you to produce or simulate authentic effects and antique finishes on wood.

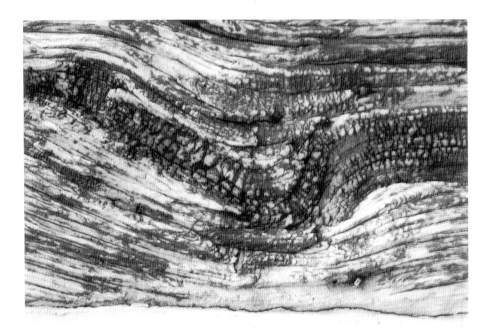

Fig 34 The natural effects of 400 years of weathering on external oak.
(This would be an ideal copy example to simulate)

Before any work can be done to the finish as far as distressing is concerned, work must first be carried out to the substrate. Flat surfaces are a give-away, and so, with the use of simple tools, distressing of the surface of the wood is paramount. Scratches are made with a piece of flint or brick, grooves are cut with and across the grain, corners are rounded off with a rasp, and splits are cut with the grain. To the artistic 'craftsman', woodworm holes are drilled in clusters, and a wire brush is used to apply minute scratches across the grain. Every wood finisher has his or her own methods of distressing wood – some use blowtorches to produce burned effects, some use acids and bleaches. The distressing must, however, be controlled otherwise it can sometimes get out of hand and look silly.

Chains, hammers or bunches of keys can also be used with great effect by slightly hammering these objects on to the surface of the wood to simulate wear. The time comes when the wood finisher has then to turn to his colours and finishes to create colour, light, shade and a suitable finish – in other words – a 'patina'.

Distressing and Other Exotic and Simulated Finishes

I have heard it said many times by antique dealers and on T.V. antique programmes that you cannot reproduce a patina. This in my opinion is nonsense – it all depends upon the skill of the operator.

Firstly, the choice of stain used is vital to the foundation of good distressing. A chemical stain is ideal, followed by wiping on bleaches in certain areas, plus the odd ink stain (real ink), followed by the tell-tale ring marks made by cups, glasses and so on, are also useful. When the stain effect has dried, the parts of the surface that stand a little proud should then be rubbed gently with worn abrasive papers (240 grade) to lighten these parts. Areas around knots should also be shaded lighter than the rest of the wood, while all corners and scratches should be rubbed in with darker pigments such as brown umber, black and a little shellac as solvent, and when still wet, a little white pigment blown into the crevices. On horizontal surfaces, a burn mark or two on the corners simulate cigarette burns, and can add to the effect.

Dulling the corners with a mixture of brown umber and black together with solvent (weak) applied with a spray gun gives a foggy effect which creates a very soft ageing look, while a few splatter spots, either white or brown, do likewise.

It is here that a finish base coat is required, either a shellac coating by hand rubber or a spray coating of a sanding sealer, simply to seal the pigments. When dry, a thin pigment wash of, say, yellow ochre and brown umber and a little shellac sealer, can be glazed over at random intervals, but not regularly. Nothing must be regular – it is so easy to colour one corner and then another, and this must be resisted. In the art of distressing, nothing must appear even or uniform. I have seen

Fig 35 Basic hand tools used in the technique of distressing wood

commercial distressing so evenly applied that every corner has exactly the same distressing markings, and it is a pitfall which it is easy to fall into.

The antique restorer or individual craftsman, working on a single item can afford to give more attention to this sort of detail than a firm involved in producing reproduction furniture. Commercial undertakings cannot afford the time to spend on distressing so, whether working on furniture or in the shop fitting industry, in shops, hotels and bars – they have to rely on the spray gun extensively with cellulose finishes to surface coat in matt or semi gloss. The spray gun is an excellent tool for the application of spatter and fog effects, which is time-consuming to do by hand, but it does tend to produce a uniform effect.

Great care must be taken to reproduce a 'patina', and shellac, oil varnishes and cellulose finishes can be used for this purpose. In the antique restoration trade, french polish is used extensively. The surface finish must be dulled down in a random manner and the use of waxes containing pumice powder, or in conjunction with wire wool (0000 type), can be utilized to produce in a few hours what genuine antique finishes have taken many years to achieve.

In commercial furniture, the effect of distressing is not to deceive the public, but to give a general appearance of age with a pleasing artistic clean finish, whilst in antique restoration work, it is necessary to harmonise new repairs. What a great many people forget is that antique furniture, like any object, needs constant servicing, so when repairs are necessary they must be carried out as soon as possible with great care and attention.

In the building industry, when new timbers are replaced to areas where they can be seen, the use of the adze is common. Most builders take great pains to distress these new woods so that they blend in and are compatible. New oak used on exterior building construction will soon 'distress' itself by simply being left to the elements. The tannic acid within the oak will soon react with the rain and weather conditions, and within a few years, blend in to the surrounding wood. However, if this is not acceptable to the client, then by using a weak solution of either washing soda or bichromate of potash mixed with water, which only stains the wood, will fill in the gap while nature takes over. What I hate to see is good quality work on hardwoods almost ruined by daubing on copious coats of modern pigmented varnishes, which on quality work looks ghastly.

Needless to say, it is not only wood that can be distressed. Leather is one example – desk tops, upholstery to seating of all kinds, and genuine leather can also be distressed. And we must not forget the distressing that takes place in picture frame manufacture and repair, and to gilded items.

To Summarize Distressing Techniques

The Antique Method

1. Prepare the substrate by distressing the wood using any tool that accomplishes the desired effect, such as a rasp, chisel, carving chisel, Stanley knife, adze (for structural wood), drill, wire brush, etc.

2. Colour the substrate with a stain of your choice, either water, oil or chemical.

3. Apply one coating of a shellac sanding sealer and allow to harden.

Distressing and Other Exotic and Simulated Finishes

Fig 37 *The wire brush roughing up the surface of new oak*

Fig 38 *Distressing new oak using a 1/8th chisel to simulate scratches*

Fig 39 *Use of a carving chisel to distress the top surface*

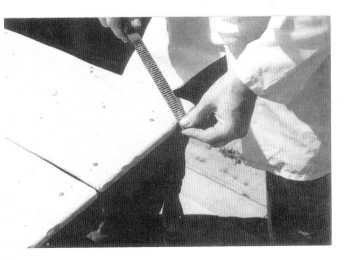

Fig 40 *Distressing the corners of new oak*

Fig 41 Finished result of distressing and chemical staining to oak

Fig 42 Simulating staining around the iron nails

Fig 43 The finished result

Distressing and Other Exotic and Simulated Finishes

4. Rub into the slightly raised areas to lighten the shades of the wood where wear would naturally take place.

5. Rub into cracks and scratches dark waxes or pigments.

6. Splatter small amounts of a lighter or darker spots of pigments at random, including ink stains.

7. Wipe over at random a light glaze coating of pigment.

8. Using an almost dry brush or mop, stipple a mixture of pigment and shellac effect in one or two corners.

9. Apply a quick coating of shellac sealer (dark), taking care not to rub off any of the above effects. This could also be achieved by dabbing the shellac on with a fad.

10. Rub with steel wool slightly into certain areas where pigments have developed too dark.

11. Using a dark wax polish, produce a fine smooth 'patina' – this can also be applied with fine steel wool.

12. In the cracks and crevices, drop in a mixture of white and brown umber – the object is to simulate dust in these areas, or you could actually use *dust* dropped onto a wet surface coating.

Distressing the Modern Way

1. Prepare the substrate to produce a fine smooth finish.

2. Using a mixed solvent stain, apply desired basic colour effect.

3. Spray on at low pressure (say 30 PSI), a thin viscosity of stainers, lacquer sealer and thinners, a fog or mist effect to the corners, taking care not to flood the surface. This is best achieved by using a colour spray gun.

4. Using a spray gun, apply splatter effects of one or two pigmented colours, such as black or brown umber, depending upon choice, which are strong colours, or if the spray gun is not fitted with a splatter attachment, using the gun spray 'through' an old twisted piece of rope or a piece of 8 oz hessian. This will give you a fine, irregular effect.

5. When all the colouring has been carried out, and has hardened, the surface coating can then be applied in the normal finishing method – a base coating or sanding sealer, followed by one or two coatings of say a cellulose pre-catalysed, or acid catalysed clear lacquer, either gloss or semi-gloss. The final finish could also be made using the new improved water-borne clear lacquer finishes, now becoming popular.

Another method of producing simulated finishes is by the printing of a wood grain foil, which is applied to a plain grained wood or manufactured wood board. This is carried out by using a special pre-etched roller that 'prints' grain on to a primed substrate.

And another method is to use an exotic veneer grain produced by a photographic technical process which is etched on to a metal roller. It is then used to reproduce vast amounts of identical copies. 'Distressed' effects can be made using the same techniques.

The method was once used for items such as radio and T.V. sets before they made them from plastic mouldings. Electronic organ and piano manufacturers now use these distressed simulated effects in great abundance, and to great effect. The only snag is that if the surface film is damaged at any time, the wood finisher must really know his job and have an artistic eye to colour in and grain by hand, in order to repair the damage. Fibreglass-type paste catalysed fillers must be used on the repairs to the substrate first.

These are the finishes of modern industrial mass-produced items we all have within our homes or offices. The superficial effect is only meant for a short time before these items are discarded; unlike good, well-made furniture or joinery, which, once made or purchased, lasts for generations. So the finish has to be commensurate with the market product.

Other decorative simulated finishes

There are a great many decorative finishes that can be achieved by pigmented colours and effects, some very simple to apply, and others quite complex. There is no end to the fantastic effects that can be achieved on wood by skilled operators. Here are a few used throughout the trade – some are traditional, some new – all are exciting to accomplish.

A Lime Finish Effect

This effect became very popular during the 1930's and went out of fashion and almost disappeared by the 70's and 80's. However, the method has been 're-discovered' and is now a very popular finish on furniture, panelling, etc. The whole idea is to use the grain of wood, mostly oak, ash, chestnut, elm, etc, and to apply a pigmented paste into the grain, wipe off and seal with a sealer. The wood must be shown, not obliterated. To achieve the effect, it can be carried out by two methods: one traditional, one modern.

The traditional way

1. Damp down the substrate to raise the grain with water.
2. Do not sand, but scrub, (using a wire brush) the fuss from the grain – the 'fuss' is the soft part of the grain – the object is to remove all soft tissue fibres out of the grain and thus open it up.
3. Slightly sand with a 240 grade abrasive paper.
4. Dust out all sanding powder and clear out the grain using a stiff brush.
5. If required, staining of the wood can be carried out, but a water stain or chemical stain only must be used. On no account use an oil based stain or the liming wax will bleed into it.

Distressing and Other Exotic and Simulated Finishes

Figs 44–45 These photographs show a sample of a new panel distressed and simulated to give the effect of age. It was commissioned to blend into the kitchen of a 17th century cottage.

118   Modern Wood Finishing Techniques

Figs 46–47   The use of distressing techniques in plastic moulding production of beams. (What appears to be wood beams are in fact made of plastic)

Distressing and Other Exotic and Simulated Finishes

6. Apply liberally a coating of the white liming wax polish which is a wax finish specially prepared for the job. Leave this on the substrate for at least fifteen minutes and wipe off all surplus wax, leaving the lime embedded into the grain and not lying on top of the substrate.

7. Leave for one hour and then apply by rubber a coating of a white or pale french polish. This can then in the normal way be brought up to either a full gloss finish, or, by using steel wool, produce a fine semi-gloss or satin finish. In using the french polish, it seals in the wax and thus will not rub off the lime polish as it is trapped within the film of shellac polish. Some very pleasing effects can be produced using this original method. This limed effect can be used on any open grained woods, but it is not suitable for close grained woods such as mahogany, beech, pine, etc. The main advantage of this effect is that the wood is still identifiable, and not obscured by the lime finish.

The Modern Way

All surface coating by spray gun

When using modern materials, liming wax pastes must not be used as they will react, particularly with the acid catalysed lacquers, so a different technique is used.

1. Damp down the substrate with water to raise the soft fibres of the grain.

2. When dry, scrub out the grain using a brass type wire brush in the direction of the grain.

3. At this stage, the substrate could be coloured by using a water or mixed solvent stain applied by rag, fad or brush or mop. *Note:* I do not recommend chemical or oil stains under acid catalysed lacquers.

4. Using a coloured paste pigment specially manufactured for liming under lacquers, apply the paste into the grain and wipe off the surplus. This will dry quite quickly.

5. Lightly sand down using 240 grade abrasive papers and dust off.

6. By spray gun, apply one coating of a pre-catalysed sanding sealer, allowing this to dry out hard, followed by one or two coatings of a clear top coat lacquer in either gloss or semi-gloss finish.

The beauty of this finish is that coloured pigments such as red liming paste can be used over a dark stain, or a yellow lime paste over a green stain on such woods as oak, for example. Thus, many combinations of wonderful effect can be achieved.

Of course, other finishes can be used to surface coat finish such as an acid catalysed lacquer or just nitro-cellulose. Stainers can be used to mix with the standard white liming paste to produce exotic colours if so desired. The advantages over the traditional way is that any wood – open or closed grained – can be limed, because of the technical advantage of the spray gun effects – almost impossible when carried out by hand.

Splatter Effect Finishes

One of the easiest finishes to produce is the splatter effect. These effects are produced by the use of a spray gun with a side air control valve and used at low pressure. To produce such an effect, first of all the substrate is coloured either by spirit staining or by using a strong pigment. This effect can also be achieved by traditional methods of flicking a paint brush loaded with a pigment colour, but this is somewhat haphazard and an outdated technique. A sanding or base coat sealer can be used to fix the colour to prevent bleeding to top surface coatings.

Using the spray gun splatter effect and a good strong contrasting colour, splatter on to the substrate until the desired effect is achieved. It should be noted that you can use one, two or more different colours if required, but a maximum of two colours is recommended. When the splatter effect has dried, the whole surface can then be surface coated with a clear finishing lacquer.

Another effect by the use of a splatter gun is by using bronze, gold or aluminium powders made up into strong mixtures and splattered on to a still-wet substrate. It should be noted that this is not advisable on vertical surfaces, but some very pleasing effects can be made as the strong metallic colour splatters create a flamboyant finish as the splats run slightly into the wet surface film.

The effect achieved by using these techniques is really endless, and a skilled operator has his or her own personal technique in producing these very exciting finishes, which are almost impossible to do by hand in comparison to that achieved by the spray gun. It should also be noted that splattering can also be applied to walls and ceilings, but emulsion paints and stainers are used here instead of wood coating lacquers, but the effect is just the same.

ICI company under their Dulux trademark, produce 'Sonata', a 'speckle effect' which is an emulsion paint (water based), containing coloured speckles. It can be applied either by hand roller or spray gun and is ideal for walls.

Pattern or Stencilling Effect

This effect is pleasing when used on furniture intended for children by partially masking a surface by using cut-outs made of either wood, metal or special waxed paper which are fixed temporarily by either pins or jigs, and then over-sprayed. When dry the cut-outs are removed to expose the shape of the pattern.

A point to mention here is that the colour must be sprayed before the pattern stencil is fixed, then over-sprayed with the contrasting colour. It must be noted that the pattern must be fixed tight up to the surface in order to obtain a clean, sharp pattern, or the colour could run under and spoil the look completely.

Sand Blast Effects

Another modern technique to age structural wood and furniture is the use of a sand or glass blaster. These can be hired, but the adherence to the safety instructions in their use must be carried out.

Distressing and Other Exotic and Simulated Finishes

These rough up the fibres of hardwoods at random, which can then be coloured by stains to great effect.

Marbled Finishes – Hand and Spray Gun Method

This is a finish that depends upon skill by the operator, and is a finish that is not suitable for mass production furnishings.

The marbling effects can be obtained by various means:

1. Spray on to a suitable substrate the chosen pigmented background colour.

2. Spray on washes at random, using very thin weak colour, through 8 oz hessian. The colours must be pale tints using white as a basic colour tinted with strainers compatible with the finishing medium lacquer. *Note:* Strainers are colour washes of white/green or white/brown, etc.

3. The main point to remember is to use a wet surface at all times so that the washes bleed into the background colours. This can be achieved by keeping the surface wet by using very thin spray washes of a thin sanding sealer using retarding thinners which slows down the drying process. The use of a colour spray gun here is an advantage. To achieve the thin line type of effect, a gull's feather can be used to etch in the hard line areas, then thinly spray over to blend in.

Another method is to use a good flood coating of a colour, then, using just the air from the spray gun, disperse the colour before it dries.

Yet another method is to apply drops of colour on to a still wet surface coating so that they bleed, and then dispersing this with the air pressure from the gun.

One or two final coatings of a clear semi-gloss lacquer is advisable so that the actual finished texture looks like marble.

It takes a little time and trouble to obtain a working technique to produce a simulated effect, but it is fascinating to do. It must be emphasized that it is important to examine in detail a piece of real marble to understand the markings in order to copy them and the colours successfully. One cannot improve on nature – only admire and copy it to the best of one's ability.

Marbling has in recent years become a very popular method of finishing cheaper joinery and furniture made from plywood, MDF, chipboard, etc, and is an excellent finish if well done.

Simulated marble can also be carried out without using a spray gun by using traditional oil-based paints and stainers, applied by mops, feathers, mottlers, liners, stipplers and any softener brushes. This can then be finished with clear oil varnishes or shellac finishes.

In the US McCloskey produce 'Glazecoat' which has endless possibilities for decorative uses such as marbling, combing, rag rolling, etc.

Pickled Pine Effect

On new pine, it is sometimes required that an effect be simulated on to the pine to make it look older and well used. This process is mainly achieved by the use of nitric acid.

Nitric acid is a very strong and powerful liquid, and *great care* must be taken at all times when handling it. It is essential to wear suitable protective clothing, eye goggles and gloves. It must also be remembered that in order to dilute the acid, the acid must be added to water – NEVER the other way round or it will spit.

The effects on pine can be fascinating. First and foremost, if a weak mixture of the acid and water is used, say, in a mixture of 1–10 by volume, the pine takes on a greyish tone. If used in greater strength a more yellow reddish tone is the result. Once the acid dries out on the pine, apply a coating of bichromate of potash or washing soda (which are alkaline substances) which react with the acid, when some pleasing coloured effects can occur. Another effect is to dab on strong nitric acid, which gives some reddish effects.

Once dry, the problem is to be wary about finishing. A wash over with a weak mixture of borax will help to neutralize the acid, and to be on the safe side, when the surface is dry, apply a thin coating of a de-waxed clear shellac sanding sealer, followed, when dry, by a finish coating of a nitrocellulose clear semi-gloss lacquer, applied by spray gun.

Acid catalysed lacquers are not recommended as a finish, in this instance.

Another use for nitric acid is to lighten old oak. It acts as a bleach, but is more positive. After the application of undiluted nitric acid upon old, dark oak, it begins very quickly to react with the wood, and, within fifteen minutes, the areas where the acid has been allowed to lie burns its way into the fibres. When cleaned off with clean water and allowed to dry, the wood takes on a much lighter colour. This method is ideal for highlighting certain areas which require colour distressing. Great care must be taken when handling this acid, as it gives off fumes when reacting on the oak, which *must not* be inhaled; so, wear a facemask in addition to the protective gear mentioned earlier.

Sulphuric Acid

This is another colourless and corrosive fluid. When diluted with water it can produce yellow to grey/green stains on pine, but these can vary. However, if used at full strength, it will produce medium to dark effects on oak and chestnut. On pine, if the acid is diluted, and then a blow-lamp is used over the surface while it is still wet, the result will be black.

The protective working gear mentioned for nitric acid above applies equally here.

Scrubbed Pine Effect

This can be achieved by first applying a thin coating of a mixed solvent stain the colour of pale brown pine. When dry, apply by spray gun a thin wash colour of white pigment, so thin that it should be called a wash coat. This can be made up of titanium white with a little yellow ochre and brown umber added to a cellulose sanding sealer and sprayed on at random. The mixture must be of a thin transparent nature and must not obliterate the grain of the wood. When dry, spray on at random clear coatings of a semi-gloss nitrocellulose lacquer to fix the whole effect. Also spray on at random clear coatings of gloss lacquer so that the effect is uneven. As in a natural pine, there are areas of shiny patches and areas which are duller. When completely dry and the film of lacquer has hardened in about two days, the effect can be enhanced still further by rubbing all over with an antique-type wax polish applied with steel wool and burnished with mutton cloth.

Antique Pine, Worn-Paint Effect

(Similar to the scrubbed pine effect, achieved mainly by hand method.)

This is achieved by applying a coat of basic button polish (french polish) and allowing it to dry. At this stage, do not flat down to a smooth finish as this will destroy the keying effect on the wood. When dry, apply, either by rag or mop, a wash of titanium white and a little yellow ochre at random, which can be fixed with a little shellac sanding sealer. The effect of this is to give that worn, painted pine effect, whilst making the pine much lighter in colour. A final coat of, say, white french polish mopped on, or a coating of a clear varnish, will provide a clear finish. Alternatively, a coating of a clear spray shellac finish applied by spray gun could be used.

1. **Scumble or Glaze Finishes**

These are transparent or semi-transparent pigmented coatings applied over a previously painted background. They can give a very pleasing effect as the two colour tones break over each other: it is like two negatives being superimposed on to one print – both images being weak. The method used is either by stippling, highlighting, or breaking up thesurface using a rag rolling or dragging action.

2. **Graining effects using scumble pigments**

Many scumble pigmented ready-made finishes are available ready for use, and need thinning with a white spirit only. In the past, this method was used for graining or simulating woods and became very popular in the 30's and 40's. It is now undergoing a revival as fine examples of simulating wood grain can be produced, mainly for church doors and public buildings. When completed the whole effect is varnished over with a gloss finish.

The tools required for these grained effects are variously manufactured special brushes such as:

a) Softener Brush   Made of pure bristle and used to give delicate and soft effects, breaking up hard areas of pigment.

b) Stippler   Made of pure bristle, it is used to pound a wet paint film, revealing the underlying colours.

c) Grainer   Made of nylon or bristle and used in woodgraining. Specially spaced bristles break up the wet pigment into finely spaced patterns, even stripes. Ideal for oak, ash or elm effects.

d) Flogger or Dragger   Made of long bristle to produce the correct dragged effect like tramlines.

e) Graining Comb   Made of rubber and can produce different effects in woodgraining.

f) Fan Pencil Overgrainer   This brush is used to apply darker grain effects on to a dry surface.

g) Mottler   This brush is used to distress and mottle graining glazes, and is ideal for mahogany graining.

h) Sword Liner   Made of fine squirrel hair, it is used to give exotic, bold lines on a dry surface.

The above mentioned brushes are just a few of the special brushes used by craftspeople skilled in the art of decorative painting such as graining, marbling, varnishing, stippling, dragging and similar finishing techniques.

Masking Effects

This is simply using a good quality masking tape to mask off one colour and then applying another colour over the other part. Another word for this is trim masking. The tape must be well stuck down on the substrate, and this will ensure a good line colour separation. The spray gun is a superb tool for this method, and some very interesting separations, using pigmented cellulose lacquers can be achieved. This method is not only used on wood, but on metal as well. For example, car or auto finishing trim work is carried out in the same way.

One point to remember is to make sure that the surface films are dry before removing the masking tape.

Far Eastern Effect

This is an invention of my own, and anyone who has been to the Far East will know what I mean. The effect is produced by spraying a pigment – first, say, blue – which must be a cellulose pigmented colour, and must be allowed to dry hard. Upon this colour is painted, say, a yellow in oil pigmented paint and allowing it to dry only semi-hard. At this stage, some of the yellow colour is rubbed or scraped off, leaving the blue colour exposed, and the yellow bleeding through.

The effect, which must look dry and dusty, is often used by garden designers and fashion photographers. Any combination of colour can be used, and the result fixed by an application of a dead matt oil varnish. The idea is to produce a surface which resembles shabby worn paint, showing under-colours.

Distressing and Other Exotic and Simulated Finishes

Carbon Effects

On structural work, the use of a blow lamp or a butane gas paint burner is ideal for simulating dark areas of carbon or charred wood. This must be carried out with care, or you may end up with a visit from the Fire Brigade. If the technique is to be used before new oak beams are to be fitted, all the better – the slight burning to the wood helps to hide the raw look of oak, but it must be treated at random. It is a good plan to douse the beam with water and allow to dry before fitting into a building.

Flock Spray Finish Effect

This is a finish which is very similar to coloured baize. The flock consists of fine pure fibres which are supplied in various colours such as green, red, purple, black, yellow, blue, etc. They are designed to be applied by a low pressure special spray gun. First, the substrate has to be prepared in the normal way to remove any flaws on the surface, and to have a coating of a coloured adhesive, oil paints or pigmented, slow drying cellulose lacquer. The colour of the adhesive medium must be compatible with the colour of the flock chosen. The flock is manufactured from cotton, silk, nylon or wool and on no account, once the flock has been sprayed on, should the surface be touched until dry, otherwise the lovely effect will be broken.

This method is ideal for the decoration of walls, or smaller items such as the interior of drawers, underneath boxes, artistic candlestick stands, bowls and a host of other uses. The beauty of this finish is that, unlike sheet baize, it will not peel off at the corners.

Another method of fixing a simulated flock or baize effect can be obtained in the form of a self-adhesive backed plastic sheet, which is simply cut to the required size and affixed to the substrate.

Sfumato Effect (fog effect)

This was first originated by Leonardo da Vinci in his painting 'Mona Lisa', and is a hazy, smoky, foggy effect which can give a subdued, non-colouristic character to any work, whether it be on wood, plaster or metal. In modern work, this is achieved by the use of the spray gun at low pressure, and can give very pleasing effects to furniture and joinery. It is used in the commercial world on furniture, giving corners of oak items that old Colonial look.

This is normally carried out using a thin, weak black/brown colour mixture, using cellulose or pre-catalysed sanding sealer lacquer, plus thinners to give the colour fog or haze, which, when dry, can be surface coated with a top coating in either gloss or semi-gloss clear lacquer.

In the decoration world, this effect can also be achieved by spraying on to walls and corners of rooms to give the pleasing sfumato effect. Emulsion paints would be used in this instance.

Collage

This is a method of taking very thin paper cut-outs, fabric, prints or photographic reproductions from good quality magazines and sticking them on to a prepared surface using the actual medium as the adhesive. For example, if you wanted to finish a small table with a gloss lacquer, but wanted to be artistic in the finish, your choice of photographs, fabrics, etc, would be stuck on to the first coating of a sanding sealer applied by spray gun, and allowed to dry hard. The edges of the photograph can be slightly chamfered, and further coatings of the sanding sealer can be applied until such time as the edges of the photograph disappear. Allow the surface to settle for at least two days, and apply further coatings of a clear finishing cellulose lacquer either in gloss or semi-gloss. This may take many coatings, until the photograph becomes part of, and is embedded into, the lacquer film.

As an alternative to sanding sealer and standard cellulose lacquer, use an acid-catalyst clear lacquer from the start. This type of lacquer has a better solid body in its make-up than basic cellulose. One modern use of this method is in the production of dining place mats and coasters. The whole intention of this technique is that the picture being used lies within the many layers of lacquer and cannot be detected by touch on the surface film. The spray gun here is ideal for this method of application.

Great attention must be paid when fixing the print or photograph in a layer of lacquer, and it is essential that no air pockets are left as they are almost impossible to eradicate. It is also advisable to make sure that the paper has expanded enough, say, five minutes, before fixing to the substrate with lacquer. General adhesives must not be used as they could react with the acid catalysed lacquers.

The finish can also be achieved by using oil resin varnish, brushed on, but due to the drying times of the medium, this method takes longer to complete.

Crackle Glaze

Water based glaze is a new effect finish manufactured by John Myland in the UK and has no flash point.

The glaze is recommended to be used under eggshell base or emulsion paint. The actual glaze is a translucent colour. The basic idea of using crackle glaze is to prepare a basecoat of eggshell or emulsion paint, leave the basecoat to dry, then the crackle glaze is brushed on in either thin or thick brush strokes according to choice. This film will break up or crack and show the basecoat coming through. It can then be varnished or sealed with any clear finish. Available in 1 litre, 2½ litre and 5 litre sizes.

Health and Safety Check List

1. Remember that nitric acid is a corrosive and oxidizing fluid so:
    a) To dilute, always add the acid to the water, NEVER the water to the acid.

b) Nitric acid can cause severe burns to human tissue, so always wear protective clothing – plastic or rubber gloves, eye goggles and face mask.

c) Do not inhale the fumes given off.

d) It can cause fire if in contact with combustible materials.

e) In case of contact with eyes or skin, rinse immediately with plenty of clean water and seek medical assistance.

f) Similar precautions are necessary when using sulphuric acid.

2. Always have good ventilation when working with lacquers, oil paints and varnishes, and wear suitable protective clothing including face masks for lacquers and cellulose thinners.

3. No smoking near highly flammable materials.

4. No food or drink to be consumed or taken into the work area.

5. Burn all waste swabs and cloths outside daily. They could ignite if left in the workshop overnight.

6. Keep all acids in a special locked cabinet out of strong sunlight.

CHAPTER THIRTEEN

Temperamental Wood Substrates

This chapter features a number of woods that are in common use which the wood finisher will encounter at some time or other in his or her working span of activity. In particular, I intend to focus on those woods that have problematical attributes, some of which are associated with the characteristic structure of the wood and others which arise following the application of certain finishing materials.

It should be remembered that not all woods will take a surface coating. Some do not under any circumstances require the application of any type of finishing material, despite claims to the contrary by manufacturers. Sometimes the public is obsessed with wanting to improve on Nature. Indeed, there are vast quantities of finishes on the market today which can be used to alter both the colour and texture of this wonderful material. There are many woods, however, such as oak, cedar, teak and iroko, which will stand and last for many years simply by being left alone.

Teak

Teak grows magnificently in tropical rain forests, and it is an exceptionally fine timber. It is resistant to attack by insects, weather and acids; it is heavy and durable and, due to its oleoresinous content, it is ideal for external use. It is also used extensively in furniture manufacture as well as for veneers. However, it is an extremely troublesome wood for the wood finisher to deal with if it is used in conjunction with modern lacquers in furniture finishing.

Any form of acid catalyst will react with the natural resins in the wood, and if a spray system is used a great deal of time would have to be spent on eliminating cissing or fish eye of the surface material. To counteract this problem, the surface should be wiped down with cellulose thinners, clear methylated spirits or anti-cissing fluid just prior to the application of the sanding sealers. Another way of counteracting cissing on teak is to spray on a heavy mist coating of the sanding sealer and leave this to harden. This forms a rough key surface for further coatings. If all else fails a very thin coating of a dewaxed shellac sanding sealer can be applied to the surface by either spray gun or fad. On no account must ordinary wax shellac (french polish) be used, as this will retard the resistance qualities of the lacquer being used.

What, then, is the ideal finish for teak? If the wood is made into a product that will spend its time outdoors, the answer is nothing; but for cosmetic reasons, a little clear mineral oil can be applied from time to time so that it keeps its colour. If the wood is to be used for interior purposes, such as furniture construction, veneers, etc, then quite a varied selection of materials can be used, such as Danish oil, shellac, cellulose, mineral oil, organic oils and wax, to name but a few. With skilled techniques, synthetic lacquers can be used, but generally this is on veneered surfaces only.

Finishing Pine Wood

Pine wood is a general term for a group of woods which includes Baltic pine, pitch pine, Scot's pine, Ponderosa pine, etc.

During the last decade or so, there has been a great demand for pine wood products in the form of furniture and joinery finishings, and it is one of the boom materials of the period. Old cottage furniture has been stripped of its many layers of paint and then coated with wax. Suddenly, waxed pine furniture has become socially acceptable, and a popular finish for areas such as kitchens, bedrooms and bathrooms. The wax and polish era arrived, and any old, badly-made piece of furniture knocked up years ago by village carpenters was brought out into the open, stripped and waxed in all its ugly glory and sold for many times its original value. Old furniture, no matter how badly made, or infested with live or dead woodworm, has become fashionable. This, together with the public's reaction against plastic laminated surface finishes, has led to a return of the 'country cottage' image, in the form of beams, joinery fittings and furniture.

A growing number of manufacturers climbed on to the band-wagon and produced finishes to satisfy the consumer; not for them the caustic soda stripping bath which was essential for producing the antique effect of old pine. What had to be achieved was a simulated 'old' effect on 'new' pine. As the market began to grow and the demand for the antique 'stripped and scrubbed pine effect' increased, certain manufacturers began to develop stains, emulsions, lacquers and acrylic varnishes for application to pine wood. Nowadays, many well-known manufacturers have a pine section in their catalogues and we see on the market pine furniture stained in mahogany, walnut and many other colours, demonstrating the versatility of this softwood. The most popular choice of finish is the pre-catalysed lacquer, which is a fast drying film that has many sheens, full details of which have been given in another section of this book.

Pine, however, suffers from one defect – that of knots. These are points where branches originally emerged from the trunk of the tree, and they are very hard and resinous compared to the surrounding soft wood. As the soft areas of the wood (which makes up the largest percentage) shrink with the passage of time, the knots remain constant. Moreover, they ooze resin. The behaviour of knots can create problems at the finishing stage. If the surface coating is oil paint, the resin actually pushes off the paint around the knot, or if a varnish surface coating is applied, the resin around the knot will bleed through the varnish film. If the knots are treated with a 'knotting' fluid, which is generally a dewaxed shellac sealer, prior to the application of any surface coating, this can prevent the problem. However, no matter how the problem is treated, they will show up sooner or later due to the structure of the wood. It is often noticeable, even on first class furniture, particularly on table tops or flat horizontal surfaces, where the movement of knots can break up the surface coating film of either lacquer or varnish. One could argue, however, that this is only a small disadvantage when compared to the positive features of this popular fast growing wood. Unlike the hardwoods, it is also more environmentally acceptable in today's climate of conservation.

In the past, pine has suffered from the application of oil solvent varnish, which in time causes it to yellow with age. The industrial chemist has arrested the problem, however, by incorporating ultra violet light absorbing compounds in surface coatings. Use of an acid catalyst lacquer instead of a pre-catalysed lacquer also assists in the anti-yellowing effect, as resins used in an acid lacquer form their own resistance against UV degradation.

There is a vast array of stains and types of lacquers available for finishing pine. The sheens that are produced can often baffle wood finishers, but we should look upon pine as a wood with its own natural sheen, in this case semi-gloss. In the past, the public's choice of sheens has been mainly gloss and it seems that every piece of furniture made for centuries has been polished gloss. Wood is not naturally gloss finished, and thank goodness for the fact that the trend today is for matt and satins.

Reclaimed Pine and Finishing

Stripping old pine with hot caustic soda is not only a lengthy process but it also causes the pine wood to darken. To overcome this problem the wood can be bleached whilst it is still wet. Once the item of furniture, door, beam, etc has been stripped and bleached, it is advisable to wash it down very carefully with free running water and allow it to dry out. If this process is not carried out meticulously, then problems can develop when finishing surface coatings are applied.

The biggest drawback with caustic soda or the hot caustic soda bath system is that it raises all the natural fibres from deep within the surface of the pine and careful sanding does little to improve the surface. To overcome this difficulty, once the item of pine has been washed and allowed to dry out thoroughly, two coatings of a pale shellac sanding sealer can be applied by spray or brush. After allowing the shellac film to harden for say 2–4 hours, the surface of the wood can be sanded smooth by using 320 grade Lubrisil silicon carbide coated papers. Using this type of paper helps to restore the surface of the wood and makes it easier to flat down, so producing a smooth, even surface. If the grain of the pine is very badly raised, a further coating of shellac may be necessary, followed by re-flatting. After dusting off the residue, give the sanding sealer a top coating of white french polish (clear) and allow it to dry. After re-flatting slightly, the surface can be waxed using a white wax polish. This polish will not darken or alter the colour of the pine in any way; white wax polish is not what it seems – it is a polish which uses bleached waxes and although it looks white in the tin, it is in actual fact a clear wax. It should not be confused with white liming wax. If pine is waxed without a shellac sealer, the wax actually darkens the pine again.

The yellowish effect of naturally ageing pine can be created by applying the sanding sealer in the same way as before, followed by a brown wax polish instead of the white. Alternatively, after sealing with a shellac sealer, apply one or two coatings of a semi-matt polyurethane or pale varnish. This is ideal where the pine is subjected to steam or water, such as in a bathroom or kitchen or anywhere else where a hard washable surface is called for.

New coatings are forever appearing on the market for finishing wood. One of these is an emulsion product manufactured in the UK by W.J. Jenkins, which is a combination of analine dyes and lead free pigments made up into an emulsion. It also contains a little ammonia to help penetration into the fibres of the wood. When dry it can be buffed to produce a smooth semi-gloss sheen and is available in many colours to suit other woods as well as pine.

If an antique finish is required, then a liming wax can be used followed by a coating of white french polish to seal the wax, and then burnishing.

For the purist, a simple coating of a clear wax polish can be applied to a light pine wood if so desired. However, it requires constant re-waxing and the change in colour will be noticeably darker.

Crystal clear acrylic varnishes are ideal for application to pine which in no way alters the colour of the wood. These new finishes are readily available in the US and UK.

With such a wide choice of finishing methods for pine it is up to the wood finisher to choose one which best suits the situation.

Iroko

This wood has become very popular in recent years with the building industry for constructing conservatories, window frames, doors, etc and for fence posts, boat building and cabinet making. Iroko grows wild as a tall savanna tree across Africa. The wood is mid-brown in colour, and is very strong, hard and durable. However, it is not the easiest of woods to work on; the grain can run wild and be ripped up quite easily when using a smoothing plane, so it is better suited to machine dressing.

The wood does not really require a surface protection coating of any kind – it will survive quite happily without man's intervention. However, if the wood is to be used externally, it can be coated with either mineral oils or synthetic oils, such as Danish oil, which do not form a skin, for cosmetic purposes only. Oil resin varnish will only last for a few months on this wood and it should not in my opinion be used; nor should boiled oil, raw linseed oil or teak oil as they do not have a great resistance against the elements.

For internal finishing, the wood can be coated with either shellac based products (french polish) or indeed synthetic lacquers and oil resin varnishes. It is all a matter of choice to suit the situation.

Jarrah

Jarrah is a hard, durable, reddish-brown wood which grows in Western Australia. It requires no finish to preserve it from either insect attack or weather conditions. Its most important characteristic, however, is its resistance to fire. It is used for structural work, such as carriage and wagon construction, flooring, rail sleepers (ideal for sleepers used in underground railway situations) and last, but not least, quality cabinet work, joinery and turning work. The wood is quite easy to work due to its straight grain and texture, and it will accept shellac polishes for cosmetic purposes on furniture. If the wood is left in an unfinished state, it darkens with age. However, jarrah does have one disadvantage – like some pines, it oozes natural resins. This is not a problem when used for structural purposes, but it can create difficulties for the wood finisher when used for interior construction, and the finishing techniques should be as for teak.

Cedar

There are many species and varieties of this wood and it is grown all over the world – Honduras cedar, Spanish cedar, Indian cedar, Nigerian cedar, Western Red cedar and the true cedars 'Cedar of Lebanon' (Solomon built his temple in Jerusalem with this wood). Most of them are classified as softwoods. The wood is well-known for its resistance to water and insect attack, as well as for its durability. It is widely used within the cabinet and joinery trades and is a delight to work with.

Today, Western Red cedar is principally used for roof shingles and greenhouse construction, due to its extreme weather resistant qualities. When used externally it requires no film-forming finish, and these should not be used under any circumstances. When the wood is used as shingles, it takes on a silvery-grey colour in time, which in no way reduces its water resistant qualities. The only cosmetic finish this wood requires is a little mineral oil. The worst finish for cedar, in my opinion, is Red Lead coloured finishes that seems so popular today, and which is produced by various manufacturers under the brand 'Red Cedar Finish'. To my mind this diabolical finish obliterates the wonderful grain and natural colour of the wood.

Cedar will accept any form of wood finish for internal purposes, such as shellac (french polish), and if the correct sealers are used, cellulose and synthetic lacquers can be applied.

Rosewood

Rosewood is one of the most beautiful of all woods and it is used for veneers on pianos, as well as for quality furniture, carving, flooring, etc. It is grown in various countries, including India, Belize, East Africa, Brazil, Madagascar, Borneo and Burma. It is a dark coloured wood which is smooth, durable, very hard and strong. The wood is very expensive but it owes its popularity to its varying colour pattern. It has a network of colour veins ranging from black to warm reddish brown, enlivened with irregular patches of pale orange, which varies according to the country of origin.

Rosewood is not, however, an easy wood to work with; it is very hard yet brittle. Due to the resin that fills the open pores, it can react with any acid catalysed lacquer and cause cissing or fish eye problems, although it does take kindly to shellac finishes, such as french polish, as can be seen on pianos of the 1930–40's period, when full gloss, rosewood veneered pianos were the vogue. Once an old, naturally-bleached or faded rosewood finish has been sanded, the true dark colour of the wood returns, which makes problems of colour matching for the furniture restorer. A point to remember here is that chemical bleaching will have very little effect upon this wood, and should not be attempted at any time.

If the wood finisher is aware of the characteristics and the many problems that can be encountered in working with this wood, a most worthy finish can ultimately be produced, which will enhance the full beauty of this wonderful wood and make it a joy to behold.

Yew

Yew is classed a softwood although it is hard and difficult to work with. It is smooth in texture and close grained, brownish to orange in colour with dark spots and used in furniture making, veneers, and for parquet flooring. As it has no resin deposits in the grain, it is ideal for table tops, either solid or veneered, but it has one problem for the wood finisher – the wood has numerous small, black open pored markings or tiny knot cavities which require filling. As the wood itself is hard it requires a thixotropic grain filler, which, after sanding, can be given a light coating of a mixed solvent stain (if required), followed by a good base coating of sanding sealer. After flatting to de-nib the sanding sealer, yew can be finished with a compatible top coating applied by spray-gun.

Oak

This is one of the most majestic of trees. There are over three hundred varieties of oak in the world today, grown mainly in Europe, Australia, Japan, USA, India, etc; the most beautiful variety of all being the English oak. It is an extremely durable and hard wood, that is resistant to weather conditions, but it is physically hard to work with. It can be used for external and internal uses, yet it is vulnerable to insect attack.

Oak presents few real difficulties to wood finishers and it has a great many advantages over other woods. When first cut it is comparatively soft, but it hardens considerably after seasoning, and it changes colour as it ages. It has a variety of uses, such as building construction, furniture, veneers, boat building, joinery and cabinet making, flooring, carving, cooperage, kitchen fixtures, etc. The wood contains a great deal of tannic acid which reacts with certain chemicals, a fact which can be used to advantage for colouring the wood with chemical stains such as bichromate of potash and ammonia (see Chemical Staining). It can also be limed to produce a light, grain-revealing finish, which is currently very fashionable.

Oak requires no surface film finish if it is to be used externally; paint or varnish will be rejected ultimately. The wood is best left in its natural state if used externally, although in time it will age and turn a silvery grey colour, which can be restored by coating with a non-film forming mineral oil if required.

As to internal uses, it will accept any form of wood finish, like shellac, french polish, varnish, wax, oil and any of the synthetic lacquers without any problems.

One further point to remember is that a black deposit will form around any iron fittings, such as bolts, screws, etc, due to the tannic acid in the wood reacting with the iron. Finally, although it is a hard, durable wood and will resist weather conditions, it is not resistant to woodworm or death-watch beetle infestation unless treated with fluids that are specifically formulated to prevent or eradicate these pests.

MDF

Medium Density Fibreboard, or MDF as it is commonly known, is the most recent man-made board. It has become very popular in all areas of the wood working trades, particularly for internal purposes, where it is used for anything from piano construction to coffee tables. It is more stable, heavier and more durable than chipboard, and is produced in sheets of varying thicknesses. It can be sawn, moulded, carved, turned and planed, and due to its resistance to bending, is used largely as a backing material for veneers.

The edges of the material can create problems for the wood finisher as they are very porous and will soak up many coats of finishing material, which can be costly. However, some manufacturers have produced a sealer which forms a skin over the rough fibres and prevents the absorption of the final finish. As the face of the boards are extremely smooth, certain finishes have difficulty in keying to the surface. This can be overcome by applying a specially prepared base of sanding surface coating, after which it will accept many types of finishes without problems.

CHAPTER FOURTEEN

Modern Wood Finishing Lacquers

In the commercial world of wood finishing, traditional polishing methods have now almost been replaced by modern chemical synthetic lacquers. The word 'lacquer' used here must not be confused with Chinese lacquer or varnish lacquer as these are completely different in their make up and have no resemblance.

Nowadays, the term is used only in connection with cellulose based surface coatings, which, combined with many other types of synthetic materials have the common descriptive label of 'modern wood finishings'. These finishes cannot be home-made and it is left to specialized manufacturers to formulate solutions of film-making compounds in volatile solvents, which dry upon evaporation. This new world of fascinating surface coatings is complex and to an extent difficult to grasp, as each 'lacquer' is different in make-up to the next, and must be treated by differing methods of application.

Today, about 90% of wood finishing is carried out commercially by using these chemically made lacquers and most furniture, joinery fittings and reproduction furniture are finished in this way. The advantages of these surface coatings and the fast drying properties suit modern production methods of finishing wood. To understand the substance that is used, it is necessary to go right back to the source of the use of cellulose, which is the foundation of the product.

In the 1930's, the first of these new finishing materials came into full commercial use and quickly ousted french polishing as a commercial finish. This was nitrocellulose or cellulose nitrate, more commonly now called 'cellulose'. The material (putting it in its simplest form) consists of nitrocellulose, resin, a plasticizer, and a solvent. The nitrocellulose imparts hardness and ability to dry, while the plasticizer improves the flexibility. The resin increases the build up of gloss, whilst the solvent allows the compound to flow and then evaporates after doing its job.

During the early part of this century, cellulose was obtained mainly from the reaction of nitric and sulphuric acids upon cotton linters or waste cotton floss, which are rich in cellulose (hence the word nitrocellulose). Through the decades, there has been a vastly increasing demand for this material as a wood finishing (and also metal finishing) film forming product, and wood pulp from trees such as the Monterey pine, spruce, Western hemlock and the newest South Africa eucalyptus (a fast growing tree) have now become very important sources of the raw materials for cellulose production. The USA and South Africa are now the two main countries which specialize in its production, supplying worldwide to major manufacturers of finishing products.

There are many variations of nitrocellulose/cellulose lacquers; some are made rather weak in viscosity, while others have a heavier, solid content of nitrocellulose, making them available for specific requirements. They can be made pigmented, gloss, semi-gloss (satin) or matt, depending upon the use to which they will be put, and can be applied to wood or metal either by dipping (for example, the makers of chisel handles dip them in a cellulose lacquer bath), or by spray gun, which is the most common form of application, or it can be hand brush coated.

One of the great advantages of cellulose is that it dries very quickly – within fifteen minutes under dry and temperature controlled conditions. It is a very popular surface coating material for furniture where the fast drying quality is the key element for such items as chairs, stools, cupboards, shelving, panels, doors, joinery trimmings, etc, or for any use where there is not a need for a full stain, heat or water resistance. It can also be used on metal substrates – e.g. on auto finishing, etc.

Another advantage is that it is classed as a reversible surface coating, which means that the surface film can be softened by its own solvent (cellulose thinners). Also interesting is the fact that the use of a pullover solvent fluid and a pullover rubber, which will be more fully explained later, can produce a full mirror gloss finish.

The surface film, when hardened, is *slightly* resistant to normal hazards such as water, mild heat, alcohol, etc, giving a far better surface due to the rapid build-up of the material, rapid sanding and good adhesion.

A further and excellent advantage with a cellulose finish is that any surface fault or damage can, at any time in its life span, be easily repaired due to its reversibility by the normal method of 'filling in' using neat cellulose, flatting, and pulling over, and in the case of a semi-gloss or matt finish, these can be flatted by steel wool. This is very difficult to undertake with other lacquer surface finishes.

Another important use of cellulose is that it can be intermixed with french polishes to make a harder surface film, and it can also be mixed with pigments and thinned down with cellulose thinners to make a wash matching colour. The use of cellulose for spraying on to vertical surfaces, which is difficult with other lacquers is also an advantage due to its fast drying properties of between 10 to 15 minutes, which can prevent running of the surface coating.

Coverage is approximately 40 square metres per 5 litre volume (40 sq yds per gallon) and any re-coating can be carried out in approximately 30 minutes after the first coating, depending upon workshop air temperature of 65–70°F, while pullovering can be carried out within the hour and at any time after, as and when required.

The abrasive papers used on this material are silicon carbide wet and dry, using grits 320 and 600 to suit the denibbing situation, using water as a lubricant. However, due to the fact that cellulose has a low flash point, and consequently a high fire risk, it should never be used on public conveyances, nor in such public places as the underground railway systems, nor should it be used on woodwork for fire surrounds where it is in close proximity with naked flames.

Another disadvantage is that upon hardening overnight, it does tend to sink, so what might appear when the final spray coat has been applied is a different surface in the cool light of dawn. This has to be contended with and allowed for in the finishing schedule, and is part of the operator's skill in anticipating this. Either a re-coat or further pulling over is the remedy.

Cellulose finishes are not as sensitive as the acid catalysed finishes and, unlike other types of lacquers, one can take short cuts. For example, while oil stains are not recommended, they can be

Modern Wood Finishing Lacquers

used provided that a dewaxed shellac sanding sealer is applied over the stain before the cellulose finish is applied; but, compatible stains and fillers should be used and the operator must keep to one supplier of the product and not intermix them with others.

A full working schedule for producing a full gloss finish using nitrocellulose (commonly called cellulose) by spray gun application

1. Prepare substrate (see chapter or preparation).
2. Fill grain (if required).
3. Stain with either water stain or compatible mixed solvent stain.
4. Apply first coating of a cellulose sanding sealer.
5. When dry, flat down to a fine smooth surface finish.
6. Apply first coating of cellulose gloss at recommended viscosity.
7. Flat down using wet and dry 400/600 grades abrasive papers and using water as lubricant (a little white soap could also be used) and the water should be soft – rainwater or distilled water is ideal.
8. Apply a second coating of cellulose gloss.
9. When hard, flat down slightly removing any nibs using 600 grade papers.
10. Using a mild pullover solvent and a pullover rubber stroke across the substrate in straight movements with the direction of the grain until a full gloss mirror finish is obtained and leave to harden off for 24 hours. (**Note:** The pullover solvent and rubber are more fully described on p. 144.)
11. This final finish can be improved still further by burnishing with a special cellulose burnishing cream. A point to watch here is that if the grain is open, then take care to colour the burnishing cream with a pigment or the deposits from the cream will fill the open grain which is then almost impossible to remove.

A full working schedule for producing a semi-gloss finish using cellulose by spray gun method of application

1. Prepare substrate.
2. Fill grain (if required).
3. Stain using a water stain or mixed solvent stain.

4. Apply first coating of a sanding sealer.

5. Flat down to a fine smooth finish.

6. Apply first coating of cellulose semi-gloss lacquer.

7. Flat down to remove nibs, etc.

8. Apply second coating of cellulose finish and allow to harden.

Note: The resultant finish may be satisfactory, or if there are any surface faults, flat down using 600 grade wet and dry abrasive papers using water as lubricant, and finish off with fine steel wool, which will improve the semi-gloss effect.

When applying any form of lacquer, either sealer or top coatings using a spray gun method of application, the first consideration is to viscosity. It is no use having the best equipment or spray plant and depositing large lumps of unatomized spray dust upon the substrate, or, in some cases, no atomization at all, due to the fact that the lacquer material is too thick.

What comes supplied from the maker of the lacquer is basically a concentrated fluid – unless specified for other types of application, such as for machine coating or brush coating. In order, therefore, to enable the operator to use the lacquer as indicated by the maker, it has to be thinned with its own special thinner (or solvent), and when dealing with modern chemical lacquers, one must only use the compatible solvent. There is no real difference here in the meanings of the terms, thinner or solvent, simply a preference of use. The purpose of thinners is to thin down the basic fluid, thus allowing it to be atomized through the spray equipment in order to deposit a lake of lacquer which reforms itself into a hard film upon a surface which is free from lumps, runs and other surface faults. The term 'viscosity' – meaning 'resistance to flow' – relates to the way in which the basic fluid has to be thinned in order for the spray system to operate, without destroying the properties of the lacquer.

Thinning any material – paint, emulsion, etc, can reduce the various factors that make the product usable with all its advantages, and this is also so with chemical lacquers – the make up of them depends upon the actual thinning of the basic fluid in such a way so as not to interfere with the characteristic qualities.

How is this carried out? First and foremost, to reiterate, the manufacturer's thinners as specified for the particular lacquer must be used. For example, in the case of an A.C. catalysed lacquer, only an A.C. Thinner must be used, and no other – each type is chemically different.

The operator must then obtain the correct viscosity for that particular material. For most small users, a good yardstick for thinning is to reduce the basic solid material with thinners by not less than 10% by volume.

On most workshop instruction sheets, the viscosity rating markings are given (e.g. viscosity 40 seconds, or 60–65 seconds, etc): This is the time which it takes for a given amount of lacquer to flow through the measuring container, and these times vary according to the maker's instructions for the lacquer being used.

The times have been very carefully worked out by the makers, so that the operator obtains the full benefit of the lacquer when thinned according to the instructions. Most gloss formulations are

Modern Wood Finishing Lacquers

worked out requiring the higher of the ratings, whilst satins require the lower. But how, and what do these rating figures mean?

Various methods are on the market nowadays for obtaining the correct viscosity, the most popular, established method is by the use of the No 4 Ford Cup. This is a container with a capacity of approximately half a pint, made of metal or plastic, with a wide neck tapering to a round hole at the base through which the lacquer can flow. (Fig. 48)

A known viscosity factor is given by the manufacturer with a pre-determined number in seconds given as a guide line. A No 4 Ford Cup is filled with unthinned lacquer and then allowed to flow out of the hole, the flow time being measured with a stopwatch or timer. This time is then compared with that given by the maker. The bulk of the lacquer can then be thinned, and a sample taken and re-timed through the No 4 Ford Cup. This process is repeated until the operator reaches the specified time rate recommended for that lacquer to obtain best results.

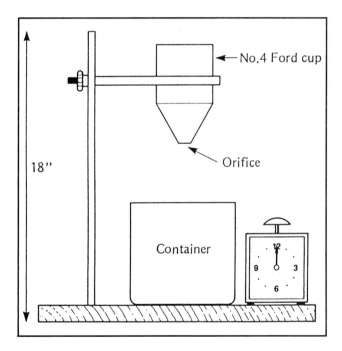

Fig 48 No. 4 Ford cup viscosity measurement device

One of the cheapest finishes is a matt, open-grained (unfilled) cellulose finish:

Schedule of work:

1. Prepare substrate.

2. Stain with a mixed solvent type.

3. Apply sanding sealer.

4. Flat down.

5. Apply one flood coating of matt clear cellulose lacquer.

If the project is a small coffee table, the whole job could be carried out within one hour, ready for hardening off and delivery to the client the following day.

In a modern finishing shop using the latest technology, this process can be carried out ready for delivery on the same day, but this dependent upon infra-red drying equipment, and the special lacquers used for this process. It does, however, give some indication of the great advantage of speed in the application of this material.

Whilst the wood finisher is not necessarily an industrial chemist, the following formula will give the reader an idea how lacquers are made up. Here is an example of a cellulose sanding sealer which typifies the complexity of modern chemical surface coatings:

A Nitrocellulose/Cellulose Sanding Sealer of reasonable quality

Nitrocellulose	6.5%
Brown Castor Oil	1.8%
Ester gum, shellac or hard resin	1.0%
Zinc Stearate Paste	3.5%
Butyl Acetate	19.1%
Butyl Alcohol	6.5%
Toluene	50.0%
Acetone	11.6%

Approximate solids 12%

It is the zinc stearate that makes the surface film of a sanding sealer easy to flat or sand down. *Note:* Sanding sealers which contain zinc stearate must not be used under *acid catalysed* lacquers.

A typical cellulose lacquer, however, is prone to after-bloom, even under stringent precautions. Therefore, ensure adequate heating and draught-proofing, and a low moisture content in the atmosphere is essential.

To Summarize

Cellulose lacquers are designed for use on furniture and joinery products where a surface film is required with fast drying properties, which also incorporates a high build in surface film, easy to pullover and which has a reasonable resistance against scratches, water and alcohol hazards. The flash point for this produce is in the range 22–32°C. There are some manufacturers who produce a High Flash range of cellulose lacquers which are much safer to use, and are very advantageous for the small wood finishing crafts. One such company in the UK marketing for this type of lacquer is Granyte Woodfinishers who specialize in industrial lacquers and enamels.

Do not forget, the lower the flash point, the more hazardous as far as fire is concerned, while the higher the flash point, the safer the product.

A Pre-catalysed Lacquer (one pack – commonly called pre-cat)

This lacquer is one of the most popular finishing products used by wood and metal finishers. The material is used by both production line operators and the small craft workshops. It is a one-pack product lacquer *containing* the acid catalyst which is added at manufacture – unlike an A.C. lacquer. When the lacquer is sprayed or coated on to a substrate, the film formation dries within 15–30 minutes at a temperature of 65°F, and with forced drying, at a temperature of 75°F, in approximately 10 minutes.

The lacquer gives a more extended time for spraying large flat surfaces than cellulose, which dries too quickly for this purpose. However, when the surface coating of a pre-catalysed lacquer dries and becomes a film by the evaporation of the solvent, it leaves the solid formation to cure by the action of the acid within the film.

In this curing stage, good ventilation is required as the odour from the curing is quite powerful, but if any work is required to be done on the film, such as pullovering, this can be accomplished within the first three hours after application. The lacquer is classed as a non-reversible finish.

On most pre-cat lacquers, the full curing time can take from 3–7 days, after which time the full advantages of the lacquer are fulfilled and the resistance qualities are at their maximum. In other words, the film is as hard as it can get, and no further odours will be emitted. The substrate can then be either packed or brought into general service or use.

It is one of the most versatile lacquers to use, since, like cellulose, it has good adhesion, and like an A.C. lacquer it has good resistance qualities which makes it an excellent all-round material to have, and can give a quality finish with variations. For example, if a full gloss finish is required, then a *gloss* surface coating is used, and this can be improved by pullovering. If a satin or semi-gloss finish is required, then a *semi-gloss* finish can be applied, and this can also be improved by pullovering. If a matt finish is required, this can come straight off the gun and left to cure, as can any of the other sheens, if required.

The great point about this lacquer is that hand finishing can greatly improve the finish, making it equal to any other finishing product, or method, such as french polishing, but having the advantages of the hazard resistant qualities.

The product is based upon a combination of urea-formaldehyde, alkyd and melamine resins, plus nitrocellulose. The pre-catalysed surface coatings consist of either clear sanding sealers, or basecoats, clear lacquers from full gloss to semi-matt and matt sheens, together with pigmented enamels, which also has the three sheen ratings. These lacquers are a great improvement on the cellulose finishes, giving high resistance in terms of abrasion, wet and dry heat, alcohol, mild acids and alkaline substances. It can be used on a variety of substrates, including wood, wood-veneers, foils, chipboard, blockboard, MDF and many other surfaces including metal. The product can be applied either by standard spray gun techniques, curtain coating, roller coating, airless, airmix (USA HVLP systems) automatic spraying systems, and also by hand brush application. The flash point is about 73°F and the product does not have the unpleasant acid odours such as with A.C. lacquer. Unlike a cellulose finish which can have a slightly yellowish tint in the finish, the pre-catalysed lacquer is extremely clear and does not have the tendency to yellow.

Full working schedule for producing a full clear gloss finish using a pre-catalysed lacquer:

Application by normal spray gun method.

1. Prepare the substrate to a smooth, dust- and fault-free condition.
2. Apply a filler compatible with this type of lacquer and sand down to a smooth finish and dust off.
3. Apply one coating of a compatible solvent stain and leave to dry.
4. Apply one coating of a pre-catalysed sanding sealer (or base coating) and leave to harden for approximately one hour. This can be less with forced drying.
5. Flat down using either 320 grade Lubrisil silicon abrasive papers, or 320 wet and dry used with soft water. Wipe off residue and allow to dry.
6. Any filling of the surface for minor faults using neat sanding sealer should be carried out at this stage, followed by flatting down any such minor filling. It is always better to use the sanding sealer than use 'imput' filler, which could be incompatible with the lacquer or even fall out of crevices, etc.
7. Apply the first coating of the pre-catalysed clear gloss lacquer, making sure the surface coating is applied evenly, and taking care to use the correct viscosity as recommended by the makers, and using a retarder thinner. This is always advisable in areas where temperature in the spraying area is variable, as it prevents blooming of the lacquer. Allow this coating to dry for at least one hour.
8. Use 600 grade silicon carbide coated papers (wet and dry) with water as a lubricant, to denib any minor faults on the surface film and wipe dry.
9. Re-coat, using the clear gloss lacquer, making sure that the whole surface is evenly covered, and allow to dry.
10. After no more than one to two hours, depending upon the ambient temperature in the spray area, the surface can now be pulled-over. This may require slight pre-denibbing or final eradication of any surface fault, such as a slight run of lacquer, etc. Use 600 grade abrasive paper with plenty of water and wipe off.
11. With a pullover rubber and a strong solvent, using straight strokes with the grain, work over

Modern Wood Finishing Lacquers

the surface until it looks like a mirror, and carry on working the rubber until all streaks are removed. (This pullovering could take 5–10 minutes or more.) The whole surface is then left to cure and no further work of any kind should be done until the full curing time has been completed. Depending upon the manufacture of the lacquer, this could be from 3–7 days.

12. After the full curing time has been met, the surface film can then, if required, be burnished using a fine burnishing cream applied with mutton cloth to complete the whole operation, after which the substrate is ready for packing or use.

Full working schedule for producing a semi-gloss finish using a pre-catalysed clear lacquer:

1. Prepare substrate to a smooth, dust- and fault-free surface.
2. Apply wood filler compatible with the lacquer, and sand down to a smooth finish.
3. Apply one coating of a mixed solvent stain and leave to dry.
4. Apply one coating of a pre-catalysed sanding sealer or base coating and leave to harden for approximately one hour. Sand down to a smooth finish.
5. Flat down using a fine silicon carbide coated paper 320 grade.
6. Adjust surface by filling any minor surface faults using neat sanding sealer, and allow to dry, re-sand and dust off.
7. Apply one good flood coating of a semi-gloss pre-catalysed lacquer thinned down to the recommended viscosity, and leave to dry and cure. The resultant surface may or may not require any further work, but if, however, there are any minor surface nibs, etc, these can be removed after one hour by using water lubricated 600 grade silicon carbide abrasive paper. Any slight abrasion due to de-nibbing can be removed by the use of 0000 steel wool. The surface can now be left to cure for the normal period.

A quality semi-gloss finish (lacquer patina effect):

Applied by standard spray gun system.

The versatility of this lacquer is such that various sheen effects can be produced. Here is one of them, which is used where a quality lacquer finish is required, well above the call upon a standard semi-gloss sheen effect.

1. Prepare the substrate to a smooth, dust-free condition.
2. Apply a wood grain filler compatible with the lacquer.
3. Apply a mixed solvent stain and leave to dry.

4. Apply one coating of a pre-catalysed sanding or base coat sealer and leave for one hour to harden.

5. Flat down using 320 grade silicon carbide coated paper.

6. Check surface for any minor faults and adjust and make good.

7. Apply first coating of the semi-gloss lacquer, making sure that the whole surface is covered evenly, and allow to dry for at least one hour.

8. Using 600 grade abrasive papers, de-nib and flat down to a smooth finish.

9. Re-coat using the same viscosity of semi-gloss lacquer and leave to dry.

10. After no more than two hours, depending upon the ambient temperature in the working area, slightly flat down again, making sure that any fault, no matter how slight is flatted down. Wipe off and dry.

11. Here the process of pullovering is applied as previously described – the object simply being to eradicate the flatted surface faults. Then, using 0000 steel wool, work on the surface until it takes on the appearance of a *glazed semi-gloss* finish. This takes some while to achieve, but the result will be rewarding and a quality finish will result. This is what can best be described as a 'lacquer patina effect'. This finish is seen on quality furniture, and will require some skill to produce.

Pullover Rubber

Mentioned within this chapter are references to a 'pullover rubber'. This is made exactly as a rubber for french polishing, but must be kept entirely separate in its own container after use.

Pullover Method

The object of pullovering is to produce a mirror gloss finish.

The method of using pullover solvents is as follows:

The surface film must first be de-nibbed before pullovering, and all dust removed. Working as if for use with french polish, charge the rubber wadding with pullover solvent; it should not be dripping, but just wet enough to well moisten the wadding. Cover the wadding with an absorbent cotton cloth, and make up into a rubber in the normal way. Start at the bottom of the panel or piece of work and work upwards by using straight strokes in the direction of the grain. The rubber must be kept moving at all times, and the process must be carried on until the mirror gloss has been produced without any surface faults. When completed, the surface must be allowed to harden off over night.

Modern Wood Finishing Lacquers

Pullover chemical solvent

A typical pullover fluid solvent is made up for a cellulose lacquer as follows:

By volume:

Special petroleum ethers	50%
Methylated spirits	25%
Butyl acetate	10%
Butyl alcohol	10%
Toluene	5%

This is a mild solution and is one of many such formulations which are made to suit either cellulose or pre-catalysed surface films.

What in fact this type of solvent does is to dissolve slightly the film of lacquer, which is 'pulled over' by the fluid on the heel of the rubber, which quickly dries as it passes over the film of lacquer. It takes practice to master the skill of this technique.

Acid Catalysed Lacquer

Commonly known as a two-pack cold lacquer or simply, A.C.

These are powerful lacquers, sometimes called cold cure lacquers. They are made with synthetic resins which do not normally dry by themselves. To harden them as a film, an acid catalyst has to be added or mixed into the lacquer just before application. Therefore, an A.C. lacquer dries partially by evaporation of the solvent, followed by polymerisation due to the presence of the acid, which reacts with the chemical components, producing polymerisation. This A.C. lacquer is classed as a non-reversible lacquer, due to the fact that the chemical reaction in curing the film cannot be dissolved in its own solvent.

The acids used as catalysts could be butyl phosphate, sulphuric or hydrochloric acids. The solid contents† of these lacquers are much higher than cellulose, so there is less sinking upon curing.

The lacquers are sometimes referred to as 'A.C. Melamine lacquers', and these, as the name implies, contain melamine, which improves the surface film, particularly in its ability to resistance against wet heat, which was once one of the snags of a standard A.C. lacquer. A disadvantage with this type of lacquer is it will not withstand strong alkalis. However, due to constant improvements by some manufacturers, some of them now are resistant against wet and dry heat. The main uses of this lacquer are on bar tops, counter tops, furniture for schools, ships, offices, hotels, etc.

The main properties are greater heat, water and alcohol resistance, which are far superior to cellulose and french polish. It has good abrasion resistance and a resistance against normal hazards, and of course, its ability to harden quickly.

---

† (approximately 45% solid content)

A full working schedule for producing a full gloss finish using an acid catalyst lacquer:

Application by standard spray gun method.

1. Prepare the substrate to a smooth, clean, dry condition. *Note:* If a wood filler is used, then this must be compatible with the lacquer.

2. Stain the wood using either a water stain or non-fade mixed solvent compatible stain.

3. Normally, no sanding sealer is necessary, however, some makers of this lacquer do supply a sealer, but *no other* type of sanding sealer should be used, especially those containing zinc stearates. (The sanding sealer, if used, must be thinned down to the stated viscosity.)

4. After flatting down, the first coating of the A.C. lacquer is applied by spray gun method. The mixing proportion by volume is normally 20 parts of lacquer to one part of acid catalyst, although this can vary. It is important that only the *recommended* amount of *catalyst* be used. *Note:* Special care should be taken when handling the catalyst as it contains acids and contact with the skin should be avoided. The amount required for the job in hand is mixed thoroughly. (The pot life of this homogenous mix is 4–6 hours, depending on the ambient temperature.) Thinning has to take place, therefore *use the correct A.C. Thinner* to adjust the viscosity to the maker's recommendation, which in this case is approximately 30 seconds. No 4 Ford Cup.

5. It is here that this lacquer differs from others. After touch drying has taken place within 15–20 minutes, the surface can be left to harden overnight for approximately 12 hours, although, in some cases, this varies between lacquers. If, however, a further coating needs to be applied, which can only be assessed by the knowledge of the operator, then this should take place while the surface is still either *semi-wet* or just touch dry. If the surface film is left for one or two hours and then re-coated, the applied lacquer can act like a stripper and ruin the surface and it can cause wrinkling. If this happens, there is no alternative but to strip back to the bare substrate, and carry out the process again. The curing area must be very well ventilated as this is the time when the odours can be very strong. After the full curing time has been allowed, the surface film can be improved if so required.

6. The whole surface can now be slightly flatted down using very fine silicon carbide coated papers in the 600 plus range of grades, using water and a little soap as lubricant. This is carried out until all nibs and any other minor surface faults are flatted.

7. After wiping off the residue of flatting, the whole surface can now be burnished, using abrasive pastes or creams specially made for A.C. lacquers, and this can be carried out either by hand or, more easily, by power polishing mops. This will develop into a full mirror gloss of quality. *Note:* Burnishing can be best achieved by using two grades of cutting pastes – medium and fine, then finally finished with a burnishing cream using mutton cloth and straight strokes with the grain.

The surface film, when fully cured will be resistant against normal hazards such as scratching, heat and stains, water, mild acids, alcohol, and not forgetting – children! For general surface cleaning, a little vinegar in warm water applied with mutton cloth is all that is required, followed by a dry cloth to remove any slight streak marks. Waxes in any form are neither required nor desirable on this surface or smearing will show. Some special dewaxed aerosol plastic cleaners can be useful for cleaning this surface film.

To Summarize

These are lacquers containing a high solid content are very hard when fully cured, and are resistant against normal hazards.

Unfortunately, they are not easy to apply, and are more suited for commercial use operations. The flash point for the clear lacquers are below 22°C/72°F, whilst primers and sealers have a higher flash point.

Due to the make-up of these lacquers, they have disadvantages, mainly in application, and full extraction systems are needed to remove the powerful odours during application and curing stages. Even when using a spray booth extraction system, the operators may require some extra form of face and eye masking.

The standard drying times can be reduced if forced drying techniques are used.

There are A.C. lacquers in enamel form, such as supplied in white and black, etc, and, as in the clear range, these can be either supplied in gloss, satin (semi-gloss) or matt, to suit the job in hand. For curtain coating, special formulations are obtainable.

One major point to remember with lacquers is that the operator must keep to one manufacturer's products right through the range, from fillers, stains, sealers, primer coats for pigmented colours or enamels, and thinners. Mixing up various makes of A.C. lacquers can lead to disaster!

Sheens

The term gloss, semi-gloss (sometimes called satin) and matt, are often used to describe a sheen finish to the lacquer: for example, a full gloss is just 'full gloss' while a semi-gloss sheen would be rated as 70% of gloss, a matt sheen would be as low as 10% of gloss.

These finishes, however, do vary in manufacture and a satin lacquer from one maker will differ greatly to a satin from another, as far as sheens are concerned.

Polyurethane Lacquers

These are cold cure, two pack lacquers – they are not acid catalysts lacquers.

Like all two pack lacquers, this product partly dries by evaporation of the solvents, and partly by the chemical reaction of a curing agent fluid containing *isocyanates*, providing a fast drying film with easy sanding qualities and excellent build. The cured non-reversible film has far superior qualities than the acid catalyst types of lacquer, such as greater resistance against wet and dry heat, alcohol, mild chemicals and excellent marking resistance.

This lacquer is recommended for use where a high degree of resistance against normal hazards is required, together with high gloss and build to substrates of industrial products made of plastic, wood, metal and man-made boards. Pigmented polyurethanes are also available in many colour shades.

As with other lacquers, different sheens are available in gloss, semi-gloss and matt, including a gloss marine exterior finish which is popular with the boat building trades. It is, however, with the furniture makers that this type of lacquer, with its many variations, is extensively used.

Flash points vary, with most of them between 22–32°C/72–90°F.

Viscosity is approximately 55 seconds No 4 cup, and the coverage can be around 8/10 square metres per litre.

The thinners used *must* be compatible with the lacquer and the thinning is generally carried out after the two parts have been mixed together. The pot life of these lacquers must be watched with care as they range from about 8 to 20 hours at normal room temperature.

When adding the solution containing isocyanates, the control of this chemical in measurement must be accurate, according to the instructions by the makers, or the resultant finish can be affected. Whereas with acid catalyst measurement there is a tolerance plus or minus to a degree or so, this is not so using a polyurethane lacquer.

The coating will be touch dry in from 10 to 30 minutes, and re-coating should be carried out 1½ to 2 hours after the first coating.

Stains and wood fillers used must be compatible with the lacquer, and the wood grain filling should be carried out first, followed by the stain; this ensures maximum adhesion. Stains and thinners which are spirit based should not be used as they may contain alcohol. The basic rule is that they must be compatible.

This is a crystal clear lacquer, unlike some of the acid catalyst types. Instead of using an acid to chemically react with the solids, here we have two basic substances – one being polyester resins in solvents, and the other resinous chemicals, which, when in contact with an isocyanate, crosslink with the resins and form urethane – hence, polyurethane. This I have put in its simplest form as it is very complex, chemically.

The mixing ratios vary as with other components of this complex lacquer; some mixing ratios are equal parts by volume, while others are two parts of one to one part of the other by volume. These compounds are not referred to as catalysts, because they are not actually acids, but are termed as either accelerators, reactors, curing agents or crosslinkers. The polyurethane lacquer is supplied in two fluid packs. Pack (A) which is a mixture of polyester resins in solvents, and Pack (B), which is a solution of a polyfunctional aromatic isocyanates. It should also be noted that some makers of polyurethane lacquers insist that the 'curing agent' should be used as soon as possible as the product has a very short shelf life. (This is mainly due to moisture.)

Another use for polyurethane lacquers is that they can be added to other products such as wood and grain fillers, where they have to be used under acid reactive lacquers acting as binding agents. The lacquer is available either for brush coating, curtain coating, and general spray application, and also by electrostatic spray systems.

The manufacture of this lacquer varies in solid formulations, one polyurethane lacquer produced for furniture may not be suitable for woods used externally; some clear external lacquers have a slight tendancy to yellow when used outdoors – yet they are all under the banner of polyurethane lacquer labels. The manufacture is complex and drying times are varied. Some have high flash points, some low – it all depends upon the use they are intended for, and to the variations of the chemical structure in manufacture.

As is my practice, I give a health and safety check list at the end of each chapter, but with this particular polyurethane lacquer, I feel that extra precautions must be emphasized. With all paints, lacquers, varnishes, etc, when using a spray system, hazards exist. However, when using a polyurethane lacquer, there are *extra hazards* arising from the use of *isocyanates* as curing agents. Unless precautions are taken, great harm can occur to the operator because of the effect on the respiratory system. Operators who use this product must be very careful to see that they are not allergic to the dangerous fumes given off when spraying and curing. Anyone who has a history of asthma should never be employed in any process involved with the use of isocyanates. This lacquer must *never* be used without proper exhaust ventilation systems. The operator must wear airline breathing apparatus, which is fitted with an efficient oil, water and fume filter to provide clean air to his visor. Even persons not actually employed with the spraying of the lacquers must be aware of the dangers if working near to the spraying area, and should be protected by suitable respirators. In the United Kingdom any waste material should be disposed of in accordance with the deposit of Poisonous Waste Act of 1972; other countries will have regulations which cover this aspect of work and readers should acquaint themselves with any such legislation and safe working practices.

A polyurethane lacquer is an industrial surface coating, and, in my opinion, should not be used unless the operator knows the full implications of this product, and has the required health and safety equipment. It is then a perfectly safe material to use, and the results are excellent for wood and metal finishing.

Polyester Lacquer

A non-reversible two-pack lacquer.

This type of lacquer is unique and different from all the other lacquers previously mentioned. It can be described as a catalyst finish, although not quite in the same context as an acid catalyst lacquer. However, polyester stands out as a finish all on its own, and is used mainly commercially throughout the wood finishing world. The piano manufacturing industry utilize this finish to their advantage, while other uses include such things as car fascia panels and trims on quality cars, kitchen and bedroom furniture, and many other mass manufactured quality products requiring a first class finish.

Polyester lacquer is different to the others, because the solids of the lacquer, on application, suffers very little from solvent loss, and therefore the lacquer can be applied much thicker, or with more depth than other lacquer finishes. In consequence, the surface film suffers little from solidification upon drying, and there is no fear of 'skinning', shrinking, or many other surface faults common with other lacquers when a thick layer of lacquer has been applied.

Polyester has a very high gloss, and looks like glass! It is a hard film and is resistant against mild heat and solvents. However, it is prone to cracking if the film is subjected to knock or indenting, and, due to this, the film can be damaged (like glass), making it almost impossible to repair. If the damage is, say, in the middle of a piano lid, the repair can be very difficult, but if the damage is on a corner or an edge, then it is possible for the surface damage to be repaired quite easily, if the operator has the knowledge and skill to undertake the process. It is these characteristics that make polyester unique amongst chemical surface coating lacquers.

Polyester can be classed generally as a two-pack type. It requires either pre-mixing with a catalyst or reactor before use, giving a very short pot life, or it can be applied as a two-pack system applied through a dual feed spray gun, giving more time for application due to the fact that the two components only meet when mixed upon the substrate and not in the guns. The ultra-violet type polyesters do not require any catalyst reactor – the reaction being completed by the irradiation of U.V. light.

The drying times of this product vary; if a wax type polyester is used – this has a very small percentage of paraffin wax incorporated, which helps to aid polymerization – a minimum of six hours is required before any flatting can be carried out, while a U.V. cured wax-free polyester cures in 10 seconds.

It must be realized, however, that a full gloss or a semi-gloss or matt finish can be achieved direct from a curtain coater, but to obtain the full glass-effect finish requires the more conventional process of flatting down the hard film of polyester, and burnishing using powered polishing mops, to produce the finish. This finish, however, is suitable for horizontal surfaces only, and it is not practical to apply polyester on vertical fixed panels, etc, for the simple reason that, due to the thick consistency of the product, it would run off before it had time to dry.

Pigmented polyester lacquers are now very popular, and these are used on modern furniture and many other products requiring colours and sheens. Polyester lacquers are available on the market today in full gloss, semi-gloss, matt and pigmented.

An extra note here is that in the burnishing stage, when using the wax type polyester, flatting is carried out in the normal way after the full drying period has been met. Burnishing edge mops are used to produce the full glass-like effect, which is the hall-mark of a polyester finish. For the larger scale production products, a system of edge mops are mounted on to a stand, comprising two mops, one coarse and one fine. The mops are specially made so as not to build up frictional heat flow. Flatting is better if sanding belts are used because if there is the slightest flaw left on the film it will show up on burnishing, so great care and attention to detail is required during this process.

Further points of interest:

The mixing ratios, which vary between makers must be carried out as per instructions, and no percentage tolerance or estimation be entertained. The 'catalyst' must be measured exactly.

Unlike other lacquers, this can be applied using a fairly low spray gun pressure of about 25–30 p.s.i. (pounds per square inch).

When spraying polyester, it is advisable to use an extraction system or face masks, and avoid any contact with the catalyst or reactor. In case of contact, wash skin with soap and water immediately.

Modern Wood Finishing Lacquers

An introduction to water-borne or acrylic lacquers

(Environmentally-safer surface coatings).

The finishing industry today is torn between two dividing lines of development. There are those manufacturers who have developed new and better acid catalyst lacquers, and are quite happy to continue using these chemical solvent products. However, there are manufacturers who have put most of their investment into *acrylic lacquers*, which is now the current trend. Many furniture manufacturers have considered the choice between U.V. cured acrylic lacquers as an alternative, bearing in mind the fast cure speed of the film so produced. A U.V. lacquer has almost a 100% solid content, and in comparison with an acid catalysed lacquer is almost double the solid content, which makes the product very advantageous.

The massive investment by major manufacturers in recent years in U.V. technolgy, and in producing chemical solvent-free lacquers with high solid content coatings, have made acrylic surface coating the product of 'today' and of the future. To many firms who process flat surface coating finishing systems of their products, the question of 'going acrylic' must be always in their minds. The only problem is that in reality, you either go for acrylics, or keep to the standard chemical lacquers of tried and tested application. The two basic solvent techniques of chemical and water systems will not intermix. This is the problem that has to be overcome, not to mention the environmental advantages of using a 'water solvent' (I use this word loosely.) At the end of the day, the finish is paramount, and the ease of application is of vital importance as far as cost/time is concerned. This is also coupled with the method of application, which I do not propose to discuss in this chapter. These new lacquers, like the standard solvent types, can be applied in just the same way – by curtain coating, by brushing, dipping or by spray gun. There is no difference in the method of surface coating with water-borne lacquers, and therefore, no new equipment need be contemplated as far as application is concerned.

An operator chooses a surface coating to suit his particular product and its anticipated life-span, and he therefore has to consider both the environmental issue and a suitable surface coating which will produce a hard film of lacquer. A good quality acrylic lacquer, therefore, has to be as good and even better than the best of the cellulose pre-catalysed or acid catalysed lacquers that have for so long been the mainstay of the wood finishing world.

It is important to remember that acrylic lacquers contain water which has a rusting effect upon any iron or iron-based implement. These lacquers and the equipment so used must be either made of plastic or stainless steel, or iron contamination can occur, and have a damaging effect upon the lacquer film. The solvent or 'thinners' is water and the addition of this must be carried out to the maker's instructions. The advantage of acrylic lacquer is that it has a higher viscosity than a normal chemical solvent lacquer.

When staining has to take place under an acrylic lacquer, special water-borne stains are used. These are designed to be a non-bleed type and are therefore completely compatible under water-borne lacquer films.

Acrylic lacquers that require a base coat or sanding sealer are formulated to give a good build and are easily sanded.

A good range of matching colours are available for use, which consists of a dye solution in water with a binder which can be overcoated with a clear acrylic lacquer.

Pigmented acrylic lacquers are available in colours such as white, black, grey, etc. These are known as 'primers' and can be finished with a compatible clear acrylic lacquer as a top covering film in the normal range of sheens.

A typical water-based acrylic lacquer

These are the lacquer of the future. They are water based finishes which are generally considered safer and easier to apply than the standard solvent based chemical lacquers. They are also more environmentally acceptable than the solvent based coatings. Due to new developments in the manufacture of lacquers, they are now available in a range of qualities for spray gun application or by curtain coating, dipping or brush coating. The material comes as a one pack lacquer in which a curing agent or crosslinker is mixed.

The one pack lacquer is used in circumstances where surface durability is less critical. These water-borne lacquers are supplied ready for use, and, unlike other lacquers, at the required viscosity, which is ideal at 55 seconds No 4 Ford Cup at 20°C/68°F. Small quantities of water may be added for other applications, but care must be taken not to over-thin. It is always best to use distilled water for the dilution. The addition of the curing agent or crosslinker imparts maximum resistance properties to the film, which withstands most surface hazards such as water, heat, alcohol, etc. The lacquer is also supplied in normal semi-gloss, satin and matt sheens, while the great advantage of this material is that water is the 'thinner' or solvent.

For the mixing of the curing agent or crosslinker, if the two part system is to be used, a mixing ratio of 50–1 (lacquer – curing agent) is usual, but this does vary between different makers. The great advantage of this material is that there is no actual pot life, even after the curing agent has been added, but if left for 20 hours or so, the curing agent will require a top up to maintain the original ratio mix.

As far as re-coating is concerned, up to four coatings can be applied, provided that each coating is allowed to harden for 30 minutes or more, depending upon ambient temperature which should be approximately 20°C/68°F.

Force drying, achieved by circulating warm air of up to 50°C/122°F will dry a coating of this lacquer in 10–12 minutes.

Working schedule for the application of an acrylic lacquer by a spray gun system

1. Prepare the substrate to a smooth, fault-free surface.

2. Stain the substrate using a water stain only.

3. Apply acrylic sanding sealer at the supplied ready-to-use viscosity.

4. When dry (approximately one hour), denib and dust down, using 320 grade silicon carbide wet and dry papers.

Modern Wood Finishing Lacquers

5. Apply top coating of the acrylic lacquer in the sheen chosen and, if required, add the curing agent or crosslinker to give maximum resistance against most hazards. (Re-coating can be carried out within two hours if required.)

6. Leave to touch dry which should be within the hour, then leave overnight to harden.

7. Flush out the spray gun system with clean water. (This must be done meticulously after each spray session.)

8. Substrate is now ready for use.

9. Some acrylic finishes can be slightly 'pulled-over', but all of them can be burnished to improve the gloss effect.

Note: The temperature in the working area must not fall below 18°C/65°F or the film formulation can be adversely affected.

The drying times can be considerably improved by the use of force drying methods, using circulating hot air and a U.V. lamp.

Mentioned throughout this chapter, when discussing the merits of various lacquers, the words 'resistance to normal hazards' are often found. All these lacquers, developed by the various manufacturers, are tested throughout their development. In the UK, a body such as the Furniture Industry Research Association (FIRA), accept sample pieces from a maker and carry out tests upon the surface finish which includes simulated situations that may happen to the product whilst in normal use. The tests are simulated in a controlled fashion to calibrate the following actions: scraping, crosscutting and indenting the lacquer surface film, and evaluating the result. Applying wet and dry heat at varying temperatures, and evaluating chemicals such as solvents (acetone), etc, spirits, tea, coffee and cold oils and fats are all used to test the durability of the surface film. When these tests have been carried out and meet the requirements with a comparison of a British Standard, the product is given approval. Hence, the manufacturer will state in its technical information that the lacquer is resistant against normal hazards. The USA, and most developed countries, have similar testing associations for this type of work.

Anyone connected with the wood finishing trade can, however, conduct their own controlled test for a lacquer surface resistance quality, by simply subjecting it to wet and dry heat, scratching, and testing everyday liquids such as spirits, cola, tea, coffee, etc., and making a detailed calibration of the results. In my own mind, there is no better way to test a surface film for resistance qualities than to produce the fully cured finish in the form of a kitchen table top, and leave it with a family with two or three ten-year-old children. Collect it six months later and *then* calibrate the results of how the surface film has withstood such (normal) hazards!

Actual normal usage of a surface coating is the only way of calibrating the lasting qualities of a surface – no matter how thoroughly a technical test and evaluation is carried out.

In outlining the various surface coatings throughout this chapter, I have endeavoured to stress the precautions which will avoid potential hazards in the use of these materials, including the acrylic finishes. The precautions are recommendations given as a general guide to a good level of commercial hygiene, and ultimate safe working for the operators. If the precautions are carried out, the materials so used are as safe as many other chemicals in daily use. With the exception of acrylic lacquer materials, all the other mentioned lacquers are covered by regulations which must be, by law, observed. The strictest attention, however, must be given to those regulations and guidelines, which in fact are for the good of everyone concerned.

When using cellulose, pre-catalysed or acid-catalysed lacquers, polyurethane, polyester, and also acrylic lacquers, great attention must be given to precautions to guard the health of the operator, or indeed, others working near to the working spray area where these materials are being used.

A check list for operators using chemical products

1. Wear protective clothing, plastic gloves, eye shields, head covering material and suitable face masks.
2. Use extraction spray booths wherever possible.
3. Fire fighting equipment must always be on hand and in working order.
4. Washing facilities must always be available.
5. Suitable hand barrier cream should be made available.
6. No smoking anywhere near or in the working area.
7. No food or drink to be consumed in the working area.
8. No electrical equipment that can spark should be used in the working area.
9. All other electrical equipment be professionally installed for areas of high fire risks.
10. Only lacquers and thinners etc, used on the work in hand are allowed in the working area, and any surplus materials must be kept in store, away from the working area.

Operators working with these materials are advised to drink plenty of fluids such as milk/tea/cocoa, etc, during the working sessions (e.g. when spraying).

Note: The special precautions for the use of polyurethane lacquers which contain isocyanites have been mentioned under that lacquer heading.

Whilst acrylic lacquers are much safer than chemical based solvent lacquers, care must be taken in the use of these water-borne surface coatings. They must be used in a well ventilated area, and, if using a spray system, face masks must be worn to avoid inhalation of the spray mist, and also contact with the skin and eyes.

CHAPTER FIFTEEN

Spray Finishing

Spray finishing as a modern method of applying a surface finish to wood, metal, plastic, brick, etc, goes back to the beginning of the 20th century, and companies were set up to develop the new technology to apply finishes to made-made products, other than by brushing.

When Henry Ford set up the first production line in 1925 to produce cars in America on a massive scale, the old time-consuming coach-painted surfaces had to go, due to the time and cost factor. He employed the latest technology of his time and finished his cars in one colour only –black – but applied by spray guns!

Today, not only cars are sprayed, but most domestic items like refrigerators, washing machines, cookers, furniture and a host of other products in daily use. Spray equipment is used for painting hospitals, railway stations, hotels and farm structures, and for applying anti-woodworm fluids, injecting foam for cavity wall insulation; the list is endless.

The whole advantage of a sprayed surface film is that it can be applied far faster than brush coating. For example, one operator using a spray gun system can cover as large an area per hour as half a dozen men using brushes; also, two spray-on coatings are equal to three coatings applied by brush, so the saving in cost and time is considerable.

One of the main advantages of a spray gun coating is that it is far superior to any form of brush coating, and surface film faults such as brush marks and variations of paint thickness do not exist with a spray coating film. A spray system can be applied to any substrate – steel, iron, plastic, wood, brick, stucco, concrete, plaster, canvas – in fact any kind of surface where a surface coating material has to be applied, and there is no better, easier or more economical way than by spray application.

The method of applying paint, lacquer and varnish by a spray gun is simply to break up a volume of fluid by compressed air – that is to atomise the fluid, and thus force it through the jets of a 'spray gun' on to a substrate, which then reforms into an even, flat, wet lake of fluid which is evenly distributed. When dry, the surface coating is as free as possible from any faults within that coating film – such as particles of dust, flies, deposits etc, which are called 'nibs'; or surplus fluid runs caught within the coating film; or 'cissing' (fish eye) (crater-like appearance resembling the moon's surface), and so on. The technique and skill is to produce a surface without these faults which is ultimately perfect in every way.

A beginner will make many mistakes at first, but with practice and perseverance, combined with good, clean, well-maintained and good-quality equipment, and with the correct mixing of the fluid that is being used (called 'viscosity' or 'resistance to flow'), will finally achieve the desired results.

Since the early days of the spray gun, much has improved; spray equipment firms such as

'Kremlin' and 'DeVilbiss', to name just two, are world leaders in this field. Sophisticated equipment such as compressors, spray systems, Airmix, Airless, robots, automatic spraying systems with four pass drying ovens, Electrostatic and Duotech and HVLP in the US* are now well established within the industrial 'sheds' of today. Yet in spite of all the modern technology, the hand-held spray gun holds an equal position in finishing wood and metal.

(* HVLP = high volume, low pressure)

Spray Finishing

The technique of spray finishing is not just for the initiated, but for anyone wanting to improve their finishing to a high standard, and a closer look at what type of equipment is required for the small craftsmen or D.I.Y. enthusiast is the first step.

A standard 'kit' for spraying consists first of all of a compressor unit. This comprises a motor – generally powered by electricity, a pump which is vaned to reduce overheating, and an air receiver. The belt-drive motor powers the pump which fills the receiver: this is a 'tank' for the compressed air, which can vary in size and is fitted with a pressure gauge and a drain cock or tap. Also fitted is a cut-out mechanism so that when the receiver is full of air, the motor cuts out and idles until such time as some of the air is used from the receiver, and then cuts in again to replace the displaced volume of used air. To put it another way, the pump, with the aid of the motor, automatically turns on and off at pre-determined pressure settings. Maximum air pressure is from 125 psi (lbs per square inch) up to 150/200 psi on larger commercial models. The drain tap on the bottom of the receiver is to allow water/oil, which is produced when air is compressed, to drain off, otherwise this water or moisture could cause problems with such surface coatings as lacquer, varnish, shellac etc.

On most small spraying units, air transformers are fitted, and these are sometimes called 'regulators' or 'clarifiers'. What this device does is to clean the air and collect moisture, and keep the air pressure regulated, so that the air is supplied to the spray gun cleanly and evenly. In so doing, no foreign matter such as rust, oil or water should clog the air lines.

No matter how good the equipment is for producing clean air for use by the spray gun, the quality of the gun is also an important factor. For the first-time user, there is only one type I would recommend, and that is the Suction Feed Gun, which works on a pressure of about 50–60 psi. The gun has a pot fitted below it, which can be of a variable capacity, to contain the surface coating fluid, which should be at the correct viscosity. The fluid is forced up the fluid tube by the compressed air which is mixed at the gun jet head to form an atomized spray jet stream. These guns are quite simple to use after a little practice, and the only fault I find is that they are quite heavy to handle.

The alternative that I prefer is the much lighter gravity feed gun. This type has the container on top of the gun, which gives more flexibility for spraying horizontal substrates such as table tops. Whichever type of gun is used, one important factor must be paramount to the user – it must be kept *clean*. Most professional sprayers know about this, for if the gun is allowed to dry containing, say, acid-hardened lacquer, then you may as well go out and buy a new one! Guns must be kept clean not only inside, but outside. I have seen many paint sprayers using guns with the coats of colours from

Spray Finishing

past work running and dripping down the sides of their guns, and they have to spend a great deal of time during spraying undoing the harm that they alone have caused.

It only takes a few minutes to flush out the working parts with the compatible solvent, and the fluid needle, tip, spring etc, can be cleaned out using a little solvent and a small brush. Dry off, apply a spot of lubricant to the working parts, wipe the outside of the container and gun, and that's it! Clean and ready for use next time.

Fig 49   Suction Feed Spray Gun

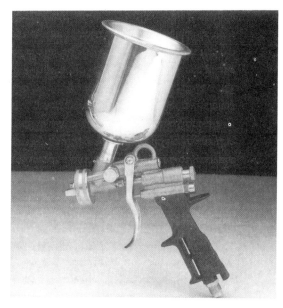

Fig 50   Gravity Feed Spray Gun

Photographs courtesy of Magnum Compressors Ltd.

There are other methods of spraying paint, varnish, lacquer etc, such as the electrical airless type, which do not work on compressed air and are very popular on the D.I.Y. market, but the compressed air type system has much more to offer to the user. Not only does it provide compressed air for the spray gun to spray such surface coatings as wood preservatives, lacquers, paints, varnishes, shellac etc, but it can also be used for such things as inflating tyres, blow-cleaning dust from electrical heaters, gas fire jets, etc, and stapling or nailing with pneumatic guns used in upholstery and elsewhere.

One of the big mistakes made by a novice in using a gun is 'speed'. The gun is subjected to treatment like a baton in the hands of the conductor of a symphony orchestra – it is waved about like a magic wand! The spray gun must be kept in a straight line towards the substrate or an uneven surface coating will result. The distance from the substrate must be uniform: if it is too near, this will result in runs – if too far away, it will give a dust bowl effect. A happy distance is based on approximately 180–230mm (7"–9"), and should give splendid results. This, however, is only possible to write about in general terms; it is with 'hands-on' practice that all will be revealed.

There are further considerations, such as the temperature in the area where spraying is being done, and, most important, the viscosity of the surface coating. It is always advisable to practice on small pieces of say, hardboard or plywood first, to get the various factors mentioned correct before you actually begin spraying your piece of furniture, car, fridge, fence or whatever. The spray gun must be used like a paint brush – think of it as one, and you will not go far wrong.

If spraying a substrate such as a table top for example, spray in a sequence which is just the same as applying paint or varnish by brush. What you are aiming for is a flat, even coating which is free from runs, drips, dusting, cissing, fish eye, and the orange-peel effect, which sadly can still be found on some production car finishes. This is the most common fault in spray finishing and is caused mainly by too heavy a spray application of the surface coating in a certain area. The heavy spraycoat builds up into an overladen peel, and also, part of the surface coating can dry unevenly.

Temperature is vital when spraying, and it is a good guide never to spray in a temperature of less than 18°C (65°F) or more than 24°C (75°F). In commercial concerns, spray booths are used and their purpose is to remove the surplus overspray fog fumes. A spray booth is simply a three sided box which can be of varying sizes, and which as at the rear either an extractor fan to take away fumes, or a curtain of water which absorbes the fumes and renders them harmless. (See chapter on Spray Booths.)

There are many firms producing spray equipment for the small craftsman or D.I.Y. enthusiast, as well as for the industrial user, and most will have a variety of models from which to choose. From the same source of supply, you should be able to get all the basic accessories you are likely to need for spraying.

The spray gun is here to stay. None of us want our cars hand brush finished, or our furniture with out-dated finishes that easily stain and mark. There is a public demand for labour-saving finishes, such as acid catalysed synthetic lacquers, which require no waxing on modern furniture and which simply require a wipe with warm water and a damp cloth. It is not just furniture that can be spray finished, but items ranging from toys, cars, cladding, machinery, fences, buildings, etc. There is no doubt that a spray kit for any small workshop is an important 'tool' to have in one's armoury.

To obtain a perfect finish on any substrate, the first priority is to choose a surface coating to protect the product from either rusting, if it is metal, or from the elements, if it is made of wood, or, if required simply as a cosmetic finish, so that the product can be properly presented to the public.

Achieving a finish that is required to meet commercial standards needs a full working knowledge by the operator. The thorough grounding through all the facets of preparation, staining, filling, bleaching, etc, and the understanding of the various finishes required for this have been dealt with in previous chapters. This knowledge must be at the finger-tips of the operator, otherwise a great deal of material, time and money can be wasted. No amount of sprayed-on surface coatings will obliterate chips, holes, cracks and other surface faults, and these must be eradicated, just as in traditional preparation, or the surface film will be full of faults. A solid understanding not only of the various surface coatings, but of the equipment itself, is also required.

The new-comer to this fascinating aspect of finishing is sometimes baffled due to the minefield of technical descriptions and terms. One of the first fundamental characteristics of spraying a fluid, be it paint, varnish, lacquer, etc, is that the fluid must be at a thickness which the spray equipment can atomise. This is termed 'viscosity' and therefore a method has to be devised which can reasonably measure this thickness. This has already been outlined in full in the previous chapter on modern

Spray Finishing

finishing lacquers, but it must be emphasized again in connection with the use of a spray gun system. To reiterate, viscosity is obtained by a measuring cup, which is a very simple method of obtaining the manufacturer's stated 'thinness' of material to give the best performance. The most popular method today is still the No 4 Ford Cup, with which it is possible, by thinning down the basic material fluid to obtain precisely the consistency required. So viscosity, or more correctly, the resistance of a fluid to flow, must be the first and foremost procedure before attempting to use a spray gun.

The full details of the various lacquers and other surface coatings have already been discussed, and this chapter is about the actual mechanics of application. Once the correct type of surface coating has been determined, the manufacturer's instructions followed, and the proper use and understanding of the technical equipment used, then, and only then, will the perfect finish be achieved.

The remaining part of this chapter will first cover the choices of equipment and their function along with the options and alternatives to meet the anticipated requirements of the workshop. There then follows a summary of the correct approach and procedures for the practical work involved in spray finishing and the methods to enable the reader to produce successful results. The moment of truth for any finisher is when the substrate and surface coating meet together – when the operator has pulled the trigger on the spray gun. This section will help you to get it right!

Types of Spray Guns

The spray gun type is identified by the position of the container or reservoir that holds the paint or other finishing liquid.

A Suction Feed Gun

A suction feed gun is of a type which has a container-cup below the gun through which a continuous stream of compressed air creates a vaccuum in the air gap, providing a syphoning action. The atmospheric pressure on the fluid in the cup forces it to the air gap of the gun, and then out in the form of atomized fluid to be reformed upon a substrate as a wet lake.

This type of gun is popular with most trades and is generally used for spraying such items as auto-finishing, furniture and metal finishings, and general wood finishing with such fluids as acrylic paints, lacquers, glues, varnishes, etc. The cups vary in their fluid capacity from 1–4 litres, but the larger cup, not surprisingly, has the disadvantage of being much heavier to hold and use.

Suction feed spray guns are made in various types to suit the particular purpose. A light industrial spray gun based on the suction feed design is ideal for use in light industry, or by craftsmen for finishing a wide selection of products. There are suction feed spray guns for fine, delicate work such as for spraying small objects, decorative work, re-touching bodywork, and also for leatherwork. The smallest of the suction feeds is a lightweight, fine airbrush type, used for painting such fine detail as strips on auto bodies, re-touching and shading work, etc, and the cups have a capacity of about ¼ litre.

Suction feed guns are made to suit the job in hand, and not always the operator. A good quality spray gun is an investment: they are expensive but it must be remembered that they are the tools of the trade, and, as with any tool, a cheap spray gun will give a cheap result – first class workmanship demands first class equipment. This type of gun can be used to spray either vertical or horizontal surfaces, and various contours, such as on auto work and furniture finishing.

A Gravity Feed Spray Gun

A gravity spray gun is very popular for finishing areas of small dimensions where the surfaces are mainly horizontal. They are ideal for the furniture finisher for such items as table tops, etc, where the air hose and large cups will not interfere with the actual spraying action. The cup for holding the paint, lacquer, etc, is placed on top of the gun and not underneath as with a suction type gun, and the beauty of this type is that various containers or cups can be used as quick change units if the occasion arises. Gravity guns cannot be used to spray ceilings, for example, but for the small workshop they are an ideal tool which will give first class results.

Spray Gun Set-Up

What is termed the set-up of a spray gun, i.e. components, consists of two basic types of atomisation. There are two different types of systems used for guns for general spraying techniques.

1. The Bleeder Type Gun
This is a gun without an air valve. The air from a small compressor flows continuously through the gun and therefore the air cannot be cut off at the gun head. This system is used where small compressors have no air-receiver (tank), or pressure controlling device. As the trigger is operated, the fluid flows out on the stream of air and, therefore, it is the trigger only which controls the flow of fluid. (This type of gun is rarely used commercially.)

2. Non-Bleeder Type Gun
This is the most-used system as the gun is fitted with an air valve which shuts off the air and the fluid flow at the same time, and, therefore, the trigger controls both air and fluid. This is how most commercial general service spray guns operate.

Spray Finishing

A glossary of terms and descriptions of the components of a spray gun

The spray gun set-up

Air Cap or Nozzle

Air caps are supplied in two basic types: the External Mix cap and the Internal Mix cap.

External Mix Cap
This type mixes and atomizes fluids with air outside of the air cap. The central hole or nozzle where the fluid leaves the gun is surrounded by a circle of air, and this assists the first stage of atomization, while other holes around the central nozzle help to provide further atomization. The horns, which have very small diameter holes, also direct a stream of air which produces a fan pattern and thus provides extra atomization.

This is the basic construction, but it can vary between manufacturers. It is used for the application of virtually all types of spray material and is most popular for the application of fast drying fluids such as cellulose lacquers. It is also used when a high quality finish is required, such as for furniture finishing.

Internal Mix Cap
This type of cap mixes air and material inside the air cap before expelling them, and is mainly used on low air pressure guns. The action is that the fluid is released from a single slot or hole in the cap, and this type of air cap is mainly used with such fluids as paint, for example, when applied to a building wall. The system is based on using low pressure and volume of air, and where the fluids are therefore slow drying, such as oil paint. It has limited use due to the fact that very little control of the spraying pattern can be achieved.

The above-stated examples are basic, and the varying types that are available on the market are complex and it is only by approaching specialized manufacturers and obtaining their full technical details, that the operator can be sure to obtain the correct type of equipment for his requirements.

The various air caps should be selected on the following factors:

1. The volume of air (psi).
2. The material feed system.
3. Type of fluid to be sprayed.
4. Size of fluid tip and air cap.
5. The size of the substrate surface to be sprayed.

These are some of the factors that require detailed choice of air cap, and fluid tip, and it is best to refer to the manufacturers' notes for the gun system being used. It must be noted that in actual practice the air cap, fluid tip and needle are all selected together as one unit as all of them combined together will affect the quality of the spray pattern and finish.

Fluid Tip

This is an important part of the anatomy of a spray gun. It is when the gun is operated that the fluid flows from the fluid tip, and therefore the rate of flow actually depends upon the diameter of the size of hole or nozzle of the 'fluid tip', and the pressure built up behind the fluid, coupled with the viscosity of the fluid. However, in the selection of a fluid tip, of which there are various sizes, consideration should be given to various factors – one being that coarse or heavy, or even fibrous fluids and materials, require large nozzle sizes so as to allow the passage of the fluid without clogging, while certain abrasive and corrosive materials must have tips made with anti-resistant and non-corrosive metals. Fluids that require atomizing at high pressures require small fluid tips.

The correct tip can be obtained by studying the makers' catalogues, which will give the standard sizes and the corresponding fluid tip opening dimensions to suit various surface coating materials.

A Spreader Adjustment

This is a valve which regulates the amount of air flowing into the air cap, and thus, by adjusting the valve, a size of spray pattern can be obtained from maximum fan pattern to a narrow round pattern.

Fluid Adjustment Valve

This controls the adjustment of the fluid needle which regulates the amount of outgoing fluid.

Trigger

This is the operating nerve centre of a spray gun. It controls the air inlet valve and fluid needle which is the only outlet for the fluid being used. The air and fluid mix is atomised under pressure once the trigger is operated.

Air Hose

There are two air hoses used on spray guns: one is red in colour, and used to carry the compressed air from the air regulator to the gun; the other is black and used in pressure feed systems to transfer fluids from the container to the gun. (The colours can vary between different makers). The air hose itself is flexible and made up of rubber, reinforced and covered with woven braid for strength.

Fluid Needle

This is a spring loaded, conical tipped needle controlled by the gun's trigger which, when operated, allows the fluid to flow; when the trigger is released, it shuts off the flow without leaking.

Pressure Feed System

Components

A pressure feed system consists of a pressure feed spray gun, a pressure feed tank and air control device, air and fluid hoses, and an air compressor.

This system consists of a large tank for the fluid to be used, and the fluid is conveyed to the spray gun through a flexible hose. The system is used where a large amount of fluid of the same mixture is to be used, and ensures that the correct sheen, colour or amount is used without variation. Additionally, a large amount of work can be done without the waste of time of re-filling small quantities of fluid to the gun system. If heavy fluids are to be used, this system is ideal due to the greater amount of fluid storage contained in one tank filling.

The principle of pressure feed spraying is the application of low air pressure on the material stored within the tank, so that it can be fed through the hose to the gun head. The air pressure is controlled by an air regulator on the lid of the tank, and a pressure gauge is provided. On some of the tanks, an air motor ensures that the fluid is kept at a constant consistency, even when the gun is not in use, and prevents solids from sinking to the bottom of the tank. On smaller tanks, a simple hand-operated agitator can be used to keep the material well mixed. Tank capacities range between 9 litres, 22.5 litres and 45 litres (2 to 10 gallons). This type of system is used to spray any type of material from the heavy solid mixtures such as cement, to normal cosmetic lacquers and paints, etc.

Remote Pressure Feed Cup

This is a separate container called a 'cup' with a mobile fluid feed system, which allows for complete freedom of movement by the operator. DeVilbiss Ransburg manufacture a 2.28 litre (½ gallon) lightweight cup, which allows for the correct balance of air and fluid on site or 'at the job' without the weight of a suction cup, compressor, and electrical equipment. (Hence the description 'remote'). It reduces operator fatigue and allows the gun to be held at any angle to work effectively. It is supplied complete with carrying handle and 1.2m (4 ft) each of flexible air and material hoses and spray gun. Also available are 'Inset Containers' for the fast and frequent change of colours and materials. They are also lightweight and easy to clean and come with carrying handles. Sizes are 9 litre, 22.5 litre and 45 litre. (In imperial this is approximately 2, 5 and 10 gallon sizes.)

The pressure feed containers can be fitted with a reciprocating air motor, which is specially designed with a three-bladed paddle agitator which rotates at a low speed, to ensure the fluid is kept constantly moving in varying directions to prevent settling.

These cups are very convenient to use in areas such as roof voids or in areas where it is difficult to use standard equipment. The beauty of these mobile cups is that they only require minimum amounts of air pressure, coupled with the fact that they have a fully adjustable pressure regulator and an accurate gauge to provide precision spray control.

Air Compressors

An air compressor is a machine designed to raise the pressure of air from normal atmospheric, which is about 14.7 pounds per square inch (psi) to a higher pressure measured in psi. Generally, compressors will give pressures up to 200 psi. For basic paint spraying, a pressure of approximately 50 psi is all that is required.

There are two basic types of compressors:

(a) The pistol compressor, which can be powered either by electricity or petrol, depending upon the design and use to which it will be put. This design elevates the incoming air pressure through the action of a reciprocating piston – as the piston moves down, air is drawn in through an inlet valve, the piston moves upwards and thus compresses the air. The compressed air is now released through an exhaust valve to the air line, and hence to the gun head.

(b) A rotary compressor, has a rotor which rotates at a low speed within an offset cylindrical housing, and blades slide in and out in precision machined slots on a film of oil. (See further full details in this chapter.)

Various makers have their own methods of manufacturing within these two basic systems, yet they all work on either of the two theories explained. For example, there are single stage piston type compressors, using two or more cylinders, or the V-type compressor using two or more cylinders but arranged at an angle to the crankcase. This shows the variety in compressor manufacturing.

Air Compressors – Types and Air Control Equipment

We should now look at the subject in more detail and at what actually provides the power for a spray gun system to operate. The air compressor can be called the power house of a spray system and the most popular system of compressing air is the 'Piston type compressor'. This is used throughout the world of paint spraying and for all forms of surface coating fluids. The range of the piston type is as varied as types of lacquer. There are the small portable compressors using a 24 litre 'receiver' tank with an air displacement of 7.5 cfm, coupled with a maximum working pressure of 125 psi and powered by an electric motor of 1.5 hp 140 VAC. This is a typical portable model, and is ideal for the home and small workshop. Top of the range compressors use a 'V' twin single phase system with an air displacement of 14 cfm (cubic feet per minute) and a maximum working pressure of 150 psi, and using a 3 hp, 240 VAC, and a receiver capacity of 150 litres. These and a whole host of similar compressors will produce clean air to the gun heads at a constant pressure.

A compressor must be purchased with care, and it is always better to refer to the manufacturer's

Fig 51 A quality, fully mobile electric 3hp Piston Air Compressor with a 150 litre capacity air receiver and 14 cfm displacement, and maximum pressure 150 psi.

Photograph courtesy of Magnum Compressors of York

catalogue for the pressure and volume requirements of the equipment using the compressed air. As a general rule, an electric piston compressor will deliver 4 cfm of usable air per hp (approximately). Compressors are either portable (with wheels) or fixed to the ground (called a stationary compressor).

A stationary compressor should be fixed on a firm, level bed and at least one foot away from a wall, which allows for the free flow of air and for easy maintenance, and also with the fly wheel facing the wall for extra safety.

The compressor should also be mounted on pads to absorb vibration, even if the base is of a solid construction. The compressor should be positioned where clean, cool dry air is available, if not then an air intake supply should be made available from the outside of the building. All electrical connections, etc, should be properly wired and a correct stop/start switch be installed.

Maintenance

Electrical or petrol-powered compressors

Always follow the manufacturer's instructions; these are basically to drain the accumulated water from the tank and the air transformer daily, check the level of oil in the crankcase, and also see that the change of oil is carried out in accordance with the instructions of the manufacturer. Blow accumulated paint/lacquer dust from the cooling fins, and other parts of the compressor. Check and clean the intake air filter (this is vital), and check the drive belt to either adjust or clean it. Generally, keep the whole unit as clean as possible, making sure that the pressure dials are wiped clean daily, so that they can be read easily.

One of the main causes of surface coating faults is the amount of water in the compressed air. When air is compressed it produces water moisture, and this accumulates within the storage tank (receiver) and air transformer. This should be cleaned out every day the compressor is in use. Most spray gun systems have extra provision for the collection of unwanted water moisture before it reaches the gun head, either by filters or by heated air.

While a compressor driven by mechanical means supplies clean air to a receiver or tank, it then has to pass through a control unit before it reaches the gun called an 'Air Transformer' – or 'Filter' or 'Regulator' – they all mean the same. This transformer removes all dirt, oil and moisture from the compressed air and regulates it. On each air transformer is an indicator dial which can be adjusted to suit the required pressure going to the gun head, or it can provide multiple air outlets to more than one spray gun system, each using a different pressure, but obtaining the constant main pressure from the receiver tank.

The main parts of an air transformer are condensers, filtering veins, an air regulator, and air pressure gauge calibated in both psi and bars, an outlet valve, and a water drain tap. Some also have a clear bowl for the visual indication of condensation level. They are also used on pressure feed tanks to accurately control the fluid pressure in the tank.

Air Regulator

This is a device for the simple reduction of main line pressure as it comes from the compressor to the gun head. It automatically maintains the required pressure as set by the operator, and maintains this with minimum fluctuations. Some air regulators are equipped with pressure gauges.

Air Condenser

This is simply an extra filter condenser which is fitted to the air line between the compressor and gun. It separates solid particles such as water/oil/dirt, etc, from the air pressure. It is an extra precaution against surface coating faults, and is fitted if required for specialized purposes.

The whole point of the condenser is to remove any type of material alien to the application of

Spray Finishing

Fig 52   The Hydrovane 25 Air Compressor equipped with a desiccant dryer and filters for ultra clean and dry compressed air.

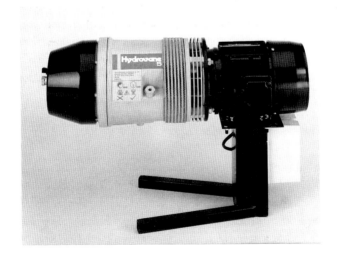

Fig 53   The Hydrovane 5 Rotary Sliding Vane Air Compressor (5 PUTs) This model is the smallest of Hydrovane's range of compressors. This 1.5 hp (1.1KW) compressor has an air output at 100 psi (7 bar) of 5.2 (2.5 l/sec) at a noise level of just 60 dBA at one metre distance. It is ideally suited for surface coatings. The clean air means less down-stream contamination, extended tool life, and less time/cost wastage – an important factor when used for chemical lacquer surface coating spraying.
Photographs by courtesy of Hydrovane Compressor Co. Ltd.

clean air supplied to the spray gun. Water moisture is the enemy of any surface coating and has to be completely eliminated from the air pressure supply and this fact cannot be emphasized enough.

The Rotor Compressor

One firm in the UK specializing in the manufacture of this type of compressor is Hydrovane Compressor Company of Redditch. They produce various types of compressors to suit all trades and industrial uses. The extensive research and development of the product have made these compressors known throughout the world. They work on a completely different principle from a piston type compressor. The major moving part is a rotor, rotating within an offset cylindrical housing. Blades slide in and out in precision machined slots on a film of oil. Air admitted is progressively compressed by a reduction of the volume of each compartment formed by the sliding blades during rotation. Oil is injected to cool the compressed air, and to seal and lubricate; it is then removed by a highly effective separation system.

A high quality compressed air is thus produced, and is ideal for operations that require particularly extra-clean air.

Specialized Spray Equipment

I have already described the three most popular forms of spray gun systems – the Suction Feed Gun, the Gravity Gun and the Pressure Feed Gun. However, there are specialized spray systems that are used by craftsmen, light industrial users and many other specialized commercial undertakings.

The first of these advanced spray gun systems, not chosen in any specific order of importance, is:

The Airless System (This system operates on pressures of 100 to 400 bars.)

Airless equipment can spray many types of material at very high pressure without using convential compressed air, and, therefore, can spray most materials much faster than other manually operated spray guns. An airless gun has an output of up to 5 litres (1 gallon) per minute, and this gives some idea of the advantage of this system when applied to large substrates. The principle of an airless system is that the motor drives a cam which in turn operates a piston immersed in oil. The slightest displacement of this piston is transmitted hydraulically to the diaphragm, which inflates and retracts. When it retracts the suction valve opens, the exhaust valve closes and the fluid chamber is filled with the fluid material (e.g. paint). As it expands, the suction valve closes and the exhaust valve opens to deliver the fluid material under pressure to the gun head.

There are two basic systems for airless spray guns – the 'Diaphragm Pump' and the 'Piston Pump.'

The Diaphragm Pumps are used mainly for the building industry and are operated by either electric or petrol motors. No air compressor is necessary with either.

The Piston Pumps are used mainly in industry and are operated by an alternating air motor which eliminates fire risks when using solvents, and very powerful pumps can therefore be introduced with this type of motor. These pumps can spray materials at high pressure *without* the aid of a compressor. Very thick coatings of viscous materials such as bitumens, and non-diluted surface coatings can be sprayed, thus cutting out the use of thinners, which is a great advantage for commercial users who save on repeated coatings. The pumps can also offset the loss of pressure when used with long hoses, and lengths of 50 to 100 metres can be used.

It must be noted that the hoses used for this system must be of a special high pressure solvent-resistant manufacture, capable of withstanding very high pressures.

Specialist manufacturers of this type of equipment include the Kremlin Spray Painting Equipment Co, of Slough, DeVilbiss, UK and USA and also Amspray in the USA, to name just three.
(Note: 1 bar = approx. 15 psi)

The Airmix System

This is a system developed by Kremlin Spray Painting Equipment and marketed under their trade mark. It was invented by this Company in 1975, and has been responsible for much improved working conditions of operators due to the dramatically improved over-spray or bounce-back conditions.

The system combines the advantages of conventional air spray and airless systems, but uses the advantages of both these systems without the disadvantages.

Standard spray systems used with compressors, and suction or gravity feed guns, etc, are all operated with fairly high air pressures from 45–80 psi, and thus a great amount of overspray is lost 'to the wind', and the cost of over-spray is a major factor in the total costing of finishing substrates, be they wood or metal. With the Airless system, the fluid fan as it leaves the tip of the gun is surrounded by two streams of air fans. The result is that the quantity and the pressure of compressed air needed to operate the system is reduced to 15–30 psi, and therefore paint or fluid fog over-spray is almost eliminated, but the resultant finish obtained from this system can be higher than a conventional spray system.

The equipment consists of an air operated pump, the Airmix gun, tip, air cap and hoses. (The system is connected to a standard air compressor set at the low pressure.)

The pump provides the fluid pressure needed to atomize the 'paint', etc. Air regulators control the outgoing fluid, and the conventional air and additional air pressure at the Airmix gun head. Airmix guns will spray most fluids such as lacquers, adhesives, paints, etc. The system is also available for automatic use, and as an extra fixing available for Hot Spray application using the Airmix system, either as a fixture or as a trolly version.

Conventional spray systems give spray fluid atomization at 30 ft/second.
Airless spray systems spray 3 ft/second.
Airmix spray at 2 ft/second.
The substantial savings of surface coating fluids can be up to 40%.
Compressed air consumption is reduced by 75% or more, and the maximum air consumption for Airmix is 5–8 cfm (cubic feet per minute). The whole system is therefore based upon a slower fluid

output droplet speed, combined with a lower pressure at the gun, which in turn reduces over-spray and bounce-back of spray fog from the substrate. If problems arise using a spray system in areas where sufficient ventilation is not always possible, then the Airmix Spray system is ideally suited.

The system is widely used throughout the car, furniture, decorating and building trades where a first-class, quality finish is required.

Spray systems produced in the USA must meet EPA guidelines (Environmental Protection Agency) for low VOC (Volatile Organic Compounds) emissions.

The Electrostatic Spray System

Electrostatic spray finishing is a very specialized method of applying a surface coating. It is based upon the law of electricity – that 'unlike' electrical charges attract each other. The system is of particular importance for spraying such items as tubes, etc, or any round surfaces where an ordinary spray system would require the operator to continually turn the object to be sprayed, thus wasting a great deal of time and material. With an electrostatic spray gun, when spraying tubes for example, only one spray position is required, and the surface coating will wrap round and apply an even coating around the tube.

Since Harold Ransburg invented the electrostatic spray painting process in 1938, the DeVilbiss Ransburg Company has been in the forefront on electrostatic development. They have the most advanced range of manual and automatic electrostatic systems currently available which are designed to provide the optimum combination of performance, reliability and versatility. Their 'Statech' guns require no bulky high voltage supply cables or complex air-driven compressors. Instead they utilize the latest state of the art 'ladder' circuitry in the actual gun to convert low voltage feed up to as much as 95 kV. *The charge is then applied to the material as it is sprayed onto the earthbed workpiece*, causing the coating to be thus attracted to the substrate, no matter how awkward the shape. This safe and simple process provides a high efficiency as it minimises over-spray, reducing spray booth maintenance and clean-up, and also provides material savings of up to 50%, which is an important factor today due to the high cost of surface coatings such as lacquer.

'Statech' guns have superb atomization and ensure constant coverage with a wide range of finishes, including the modern water-borne acrylic surface coatings.

The range of guns vary according to the particular job for which they are required. They provide a high film build on such items as durable domestic products, and they are ideally suited for spraying production line or on site furniture re-finishing to all kinds of objects.

One very popular model is the No.2 Process Hand Gun and considered by many contractors and on-site finishers as one of the most efficient to date. Utilizing pure electrostatic force, the gun's slow rotation bell delivers a soft, finely atomized spray finish with virtually no over-spray and claims up to 95% transfer-efficiency, thus saving on wastage and clean-up time. Its simple operation and portable design is a great feature of the gun, and it requires no *external* compressed air supply.

Spray Finishing

DeVilmix Air Assisted Airless System

This system is based on the combination of two finishing technologies that DeVilbiss have developed: the DeVilbiss Air Assisted Airless Spray System, which combines the application speed and simplicity of airless spraying with the control and quality finish provided by air atomized systems. The result is a thoroughly versatile finishing system which produces a controlled spray pattern with low turbulence, minimum overspray, excellent finish and reduced material use.

The DeVilmix system gives the ability to apply solvent-based, high solids, water-based coatings, polyurethanes, stains sealers, etc, to a wide range of substrates such as wood, metal, plastics, fabrics, etc. It provides a choice of spray outfits to suit specific application requirements. A selection of pump mountings combined with a choice of single or duel operator units, plus the option of a heated system. The system can also be incorporated into any production line in conjunction with high performance automatic spray guns.

Mechanically, the system is powered by pneumatic 14:1 and 18:1 ratio pumps which have been developed for continuous operation with noise levels below 80 dB (A).

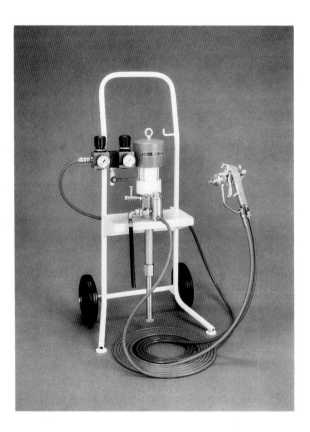

Fig 54   DeVilbiss Ransburg air assisted airless portable spray system
Photo courtesy of DeVilbiss Ransburg

To summarize, the advantages are:

Great material savings,

Better environment for the operator,

Significantly reduced noise levels,

Improved penetration in corners and recesses (e.g. drawers, etc.),

Less demand in air supply,

Reduced over-spray, bounce-back spray,

Faster film build – reduced solvent use,

Lower booth maintenance.

This system of applying surface coatings is greatly used in the furniture finishing industry where lacquers/chemical solvents and water-borne finishes, stains, sealers, etc, are used.

Fig 55 Air Assisted Airless System in use
Photo courtesy of DeVilbiss Ransburg

Spray Finishing

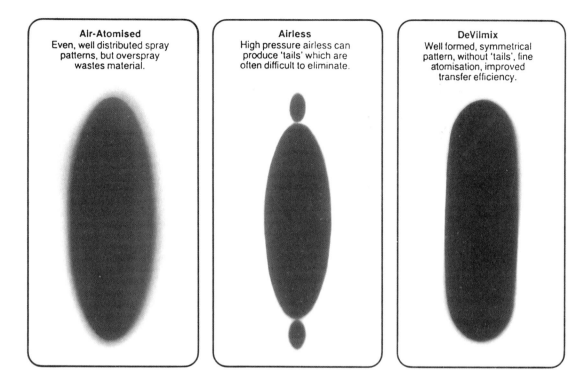

Fig 56 The Devilmix benefit of combining two finishing technologies

DeVilbiss Ransburg 'Duo Tech' System

This is a revolutionary coating system recently designed to meet increasingly strict pollution control regulations both in the UK and USA. The system combines the dual technologies of high volumes of low pressure air (around 10 psi at the air cap) resulting in higher transfer efficiencies, improved over-spray quality and reduced environmental emissions. DeVilbiss has provided industry with a genuine production spray finishing alternative that will meet new legislation.

Duo Tech Air Conversion Unit provides a regulated source of air in high volumes, and new guns have been developed to utilize the system. The concept of this system is that air can, if required, be heated. Both heated and unheated air conversion units are available, and each are capable of supplying one or two spray guns as required. The heated conversion unit is designed for continuous operation and is capable of fast warm-up and constant air temperature. The manufacturer also has a complete range of Duo Tech systems and equipment which have been designed to integrate easily with existing finishing operations.

The equipment consists of two major components: a wall-mounted air conversion unit which can be installed in any main compressor line, and a choice of manual or automatic spray guns developed to operate with high volumes of heated air at pressures up to 0.7 bar (*10 psi*). The system operates from normal conventional fluid feed systems without any adaption.

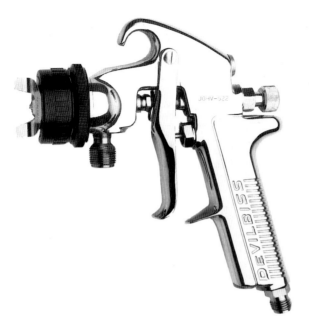

*Fig 57   The latest 'Duotech' HVLP spray gun
Photo courtesy of DeVilbiss Ransburg*

*Fig 58   'Duotech' HVLP spray gun used on kitchen
cabinet doors*

Spray Finishing

Fig 59 '*Duotech*' *applying lacquer to William Bartlet* '*Strongbow*' *furniture.*
Photo courtesy of DeVilbiss Ransburg.

Heated Application Systems or Hot Spray Application

There have been many systems in the past to improve on the application of surface coatings applied to substrates. Some methods have been to heat up the fluid before transfer, which reduces the solvents required to atomize, thus providing a thicker surface film. It must be understood that the more solvent added to a lacquer, for instance, the more obvious the disadvantages, such as loss of gloss and decrease in the depth of coating film. (This will probably result in extra applications of surface coatings, so adding to the cost/time factor.) When surface coatings such as paint, lacquers, etc, are heated, their viscosity is considerably reduced. By adding a heater unit system to the spray equipment, the control over the fluid viscosity, whatever the atmospheric conditions, are considerably improved and a more even flow and quality surface film results. The great advantage of heating a surface coating before application is that a much lower pressure is used to atomize the fluid, thus a great reduction in psi from 50 to as low as 20/25 psi, and the wasteful fog effect, or rebound off the substrate, is greatly eliminated, coupled with the fact that the operator works in a healthier working area. It must also be remembered that the actual savings on surface coating materials and solvents are thereby reduced.

However, this was the basic technology years ago, and various manufacturers produced equipment that heated the surface coating fluid before transmission. Due to new technology and development of this system, heating of fluids has now been greatly improved by actually heating the *air* that flows in transmission from the gun, but not the fluid directly.

With the 'Duo Tech' system, the added advantage is not just using low pressure air, but, additionally, heating the atomized air. The air is not just warm but heated to a manually set temperature which can greatly improve the transfer efficiency and the quality of the finish.

Using the Duo Tech conversion units, the output temperature can be regulated up to 115°C (240°F) to suit the material that is to be applied by the operator.

This system allows the transfer of the material to the substrate with the advantage of a faster solvent flash-off period, minimising blushing and blooming, and giving a better gloss finish due to the faster rate of drying, thus reducing the risk of the entrapment of dust, dirt, insects, etc, on the semi-wet drying film.

Other methods of Hot spray Application such as Kremlin's, utilizes their standad Airmix system by the addition of a 1.5kW heater and circulation unit, used in conjunction with the Airmix system, either as a wall-mounted unit or as a trolley version. This unit maintains a constant temperature for the uniform quality of the material being transmitted.

Whichever system is used, the advantages of a hot spray application system are:
Constant quality of finish,
Less over-spray,
Less pollution,
Savings on solvents,
Less surface faults,
Shorter flash-off period,
Better working conditions for the operator,
Savings of lacquers, etc.,

Spray Finishing

Faster solvent evaporation,
Thicker coating film.

Spraying Techniques

The equipment which I have described in detail of manual spray guns (with the slight exception of the electrostatic spray gun) all depend upon the skill of the operator.

Spray finishing is an art; and the knowledge and use of the equipment, how it operates and what it can achieve is all part of that craft. The finest equipment that money can buy is useless in the hands of a novice, and it is, therefore, of paramount importance that the operator knows how to control, use and maintain the equipment which is available.

This book has emphasized the basic understanding of all manner of equipment and materials, which ultimately leads to the understanding of the whole range of finishes. What makes a good spray finisher is not just the knowledge of how to apply a surface finish using a spray gun, but how to improve on the finish each and every time. After all, it is the finish that helps to sell a car – it is the finish that enhances a piece of furniture – and it is the finish that is supreme on any man-made object. The public may not understand how a piece of furniture is made, or how many hours of work that it has taken to produce it; they look at the finish, the colour and the texture. The finisher has the last say in a long production line, and if the finish is not up to a high quality, then the sale of the product is in jeopardy.

In the foreword of this chapter, I gave an introduction to the basic method of spray finishing and later, described the whole range of spray gun systems. All of them, other than automated or machine coating, depend on the same method of application, and this must be mastered by the operator. It is a manually operated method of applying a finish to a substrate, no matter which system is employed.

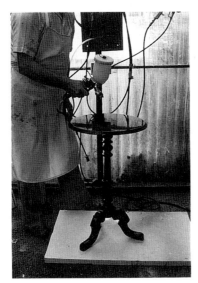

Fig 60 Spraying a reproduction table using a gravity feed spray gun

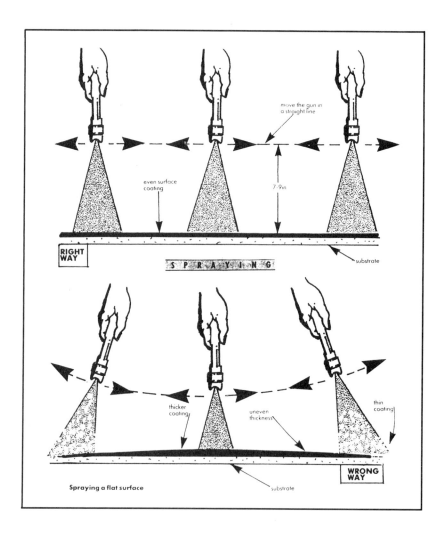

Fig 61 Correct and incorrect spraying technique

The correct method of using a spray gun

Any type of spray gun, with the exception of an electrostatic type is held at right angles to the substrate, and it must feel balanced in the hand. The distance for correct spraying technique is approximately 6–8 inches (15–20 cms) from the air cap to the surface being sprayed. All edges must by sprayed first, including inside and outside corners, followed by the rest of the main area. The speed of each successive stroke must be constant so as to maintain a uniform thickness of material. Correct triggering of the gun is also important, and is an essential part of the passing stroke technique. The operator must use the gun with great care and any trigger action must be applied smoothly. The spray pattern must be controlled so that the surface coating is blended with each passing stroke. It is here that practice in the technique of spray finishing cannot be taught by reading, and only a guide-line can be given. Like french polishing, spray finishing comes with a great deal of practice, not just in using the gun but knowing the surface coating and how it performs, such as in evaporation – the run of the coating, the speed of dry-off period, the speed of pass-strokes, etc, are all important factors in applying a surface coating. If the speed is too fast, the result is a dust bowl effect, too slow and you have a running lake of material. To obtain an even coverage, no matter what surface coating is being used, the overlap coverage should be approximately 40–50%. The gun-pass over any substrate must be in a straight line, which ensures an even coverage of material. Here again, it is practice that makes perfect, and I say again, it is the understanding of the equipment and the material that is of paramount importance. One main factor is the viscosity of the material – if it is incorrect then the finish will be incorrect, and the result will be obvious.

It is here that I must emphasize the cleanliness of equipment. The spray gun, no matter how it is used or how often, must be kept clinically clean. After each use the whole gun must be stripped down and cleaned with the compatible solvent or thinners. The outside of the gun must not show the slightest evidence of paint, etc, from previous use, for a dirty gun will give a poor result no matter what quality it is. A good operator praises his gun, while an operator who produces a poor result blames his gun! 'Cleanliness leads to quality' should be written above all spray booths!

When a spray gun system has been set up and before any actual spraying is carried out on the substrate, a test piece must be carried out on either a flat piece of plywood or hard paper in order that the gun can be adjusted for correct air pressure intake, and for the outgoing atomization and pattern of fluid, and correct viscosity, etc.

Common Surface Coating Faults

The first of these faults which show up on a substrate being sprayed is:

Dry spray
This is basically due to the gun being held too far from the substrate, and the effect is like a dust bowl; or, alternatively, it can be caused by too high an air pressure through the gun, which will give the same effect. Again, the viscosity may be incorrect and too thick. The adjustment of air pressure and correct viscosity, together with better hand control of the gun will correct this fault.

Cissing or Cratering (sometimes known as Fish Eye)
This is an operator's nightmare: the surface, just after being sprayed and when it is still in a wet condition, becomes like a close-up of the moon's surface and starts to break into craters. It can be caused by a variety of different factors: by oil or water in the air line to the gun head; or by minute particles of wax or silicones, either in the air or on the substrate, particularly when the substrate has been stripped; if ammonia has been used in certain water stains; or even if minute particles from dry silicon carbide abrasive papers (the lubricated talc), which can be deposited in the open grain of the wood substrate, are present. Other chemicals or odours from outside the workshop area can foul the air intake duct, or it can be caused by the actual substrate itself. (See chapter on temperamental substrates.)

To eradicate this problem, make sure that the equipment, the air and the substrate is clean. The use of an 'anti cissing agent', such as Bevloid, a flow agent (silicone), is helpful. (Strange as it may seem, silicones are added to counteract this problem.) Add 10% of this fluid as part of the standard viscosity. If there are small areas of cissing (fish eye) on a wet film, they must be left to semi-evaporate and the cissing craters can be spot filled using a tooth pick by dropping a small amount of the surface coating being used into each crater; when semi-dry the whole substrate can be re-sprayed. In cases where great difficulty is encountered, the whole substrate is left to dry and completely flatted down, then re-sprayed. In some cases it is quicker to strip and start again – there is never any one specific action – it all depends upon the skill of the operator in knowing what to do in these circumstances.

Blooming or Haze

The cellulose lacquer films are prone to this effect, particularly the full gloss finishes. The haze effect can occur even after a few days when the surface film has completely dried and hardened. The condition is caused by a combination of factors such as various compounds used in the make-up of the material being used – resins which are not quite compatible with the atmospheric conditions where they are being sprayed, or even where non-compatible thinners are being used.

To prevent blooming, choose a supplier of lacquer/paint, etc, that in your experience will not produce this effect. To eradicate the problem of bloom, the use of an anti-haze cream is of some help, and this consists of a formula made up as follows:

Equal parts by volume of

methylated spirits (alcohol),

white spirit or turpentine substitute (mineral spirit),

raw linseed oil,

acetic acid (white vinegar),

soft water (rain or distilled),

2 parts french chalk.

This mixture can be stored in a glass or plastic container. Shake the contents before applying to the area using mutton cloth, and burnish off the bloom. This mixture must not be used on open-grained substrates as the chalk will show up in the grain, although an added colouring pigment to suit the colour of the substrate can be used here if required.

Blushing

This has the appearance of blooming and the two are sometimes confused, but blushing occurs while the surface is still in the *wet* stage. This blushing or whitening of the wet film can occur in paint, lacquer, varnish, or enamels, and it is caused by the sudden lowering of the temperature in the spray area as the solvents evaporate from the film. Also, excess humidity in the air can cause this problem whilst the film is still touch drying off. To eradicate this fault a retarder thinner should be added to the standard thinner. Retarder thinners contain butyl alcohol-butyl acetate- or exlene. They should, however, be used with care or the surface film may dry in patches and the use of a pullover rubber may be required to restore the surface film to normal. A retarder thinner, as the word implies, slows down the evaporation of solvents, and here again, it is the skill of the operator

Spray Finishing

in knowing what action to take in these circumstances. Some makers of retarder thinners supply it to be used instead of standard thinners, depending upon which type of lacquer is being used.

Fig 62 Retarder Thinners which slows down the evaporation of solvents when added to a lacquer, and avoids 'blushing'.

Fat or Rolled Edges
This is a common fault when the edges of a substrate show a ridge of paint, lacquer, etc, and careful use of the spray pattern is the answer to this problem. To eradicate the fault allow the film to harden and flat down using a wet and dry abrasive paper (320 grade) with water as the lubricant, then re-spray over. Sometimes when spraying a flat substrate, make sure that the level is perfect or the wet film can **slide** or **run** slightly to one side. If this occurs it is best to wipe off and start again, whilst the surface is still wet.

Orange Peel
This is a sprayed coating that dries and has a texture resembling the surface of an orange. The problem can be caused by the fluid being sprayed too thickly and thus preventing it from flowing; or, alternatively, the fluid is being sprayed at too low a pressure; or, the gun could be held too near the substrate. Any of these factors could cause this problem. To eliminate the possible causes check the viscosity of the fluid being sprayed, and the air pressure from the receiver and use a better spray technique. If the fault is only a slight orange peel effect in patches, then leave to harden, then flat down in the normal way and re-spray.

Chilling
Chilling is caused by excess moisture in the atmosphere condensing on to a dry film of surface coating. The result can cause a slight dulling if a gloss finish has been used, or even a slight glossing if a matt finish has been used – one cannot win on this factor. The remedy is simple – avoid sudden draughts in the spraying area or curing area; improve the dry heat in these areas; and clean out the air line moisture filter.

Nibs
This is the enemy of all wood finishers. These small particles of dust, hair, fibres from clothing, dandruff, flies, or the residue of sanding material waste sometimes invisible to the eye, which, when the surface coating is applied, show up as very small raised pin heads within the film of surface coating. To eradicate, make sure that the substrate is clean and as dust-free as possible. This can be helped by using a **Tak Rag** which is a piece of rag with an impregnated, slightly sticky coating, which when dusted along the substrate picks up dust that the eye cannot always see. One point to remember here is that you should not rub hard on the substrate or a deposit from the tak rag itself can cause a surface film fault when using some acid catalysed lacquers. To remove the nibs, the film must be dry and you can then use either wet or dry abrasive papers, such as 400/600 grades, with a little white soap and soft water; or use the dry type of lubricated abrasive papers which will also remove the nibs.

When producing a full gloss finish, the nibs must be very carefully removed as they will ultimately show up on the film, while with a matt finish they are not so noticeable.

Printing or Flooding

This is when a small area of the substrate is still showing signs of being wet after a normal period of drying time has been allowed. This is sometimes most apparent when using semi-gloss finishes. This problem is caused by very small areas of the surface coating not allowing the solvents to evaporate. This can be caused by the use of certain pigments for colouring certain areas or it could be that moisture in patches has been present upon the substrate, or the surface coating has been applied in too thick an application. To eradicate, either apply a little extra dry heat from the heating system such as a heated blower or, with care, pull-over the whole surface film and when dry, re-spray.

Rucking

This looks as if the surface has had a chemical stripping compound used on it. It occurs when applying a second coating of lacquer when the first coat has not completely hardened, or has not completely polymerised into a film. The solvent in the second coating will lift off the first and react like a stripping compound. Alternatively, it can occur when using any of the acid hardened lacquers, if the second coating has not been applied within the time-scale stated by the manufacturer. To rectify, either strip off down to the bare substrate and start again, or allow to completely harden, flat down and flood coat using the top surface coating.

Another cause of rucking is due to the use of pigmented stoppers which may not be compatible with the lacquers being used and thus areas where the stoppers are used show up with this fault.

Special Effects using a Spray Gun System

With a standard spray gun system it is possible to achieve various effects not possible by traditional hand methods. These basic effects can be carried out by decorators, craftsmen or women, wood finishers, artists, or anyone who has studied the techniques of the spray gun and artistic colour work. The basic effects that can be produced are as follows:

Spattering

Spattering is a process of applying paint/colour pigment, etc, in the form of tiny spots. Normally, fast drying mediums are used for this process to avoid runs, so pigmented cellulose surface coatings are ideal for this effect. A low pressure is required for the process and some guns have an in-built attachment to produce a spatter effect which is simply to spray uniform spots from the gun. Various colours can be used to obtain very pleasing patterns by using multi-tone overspray colours, which, on such substrates as walls, panels, and ceilings, can be very effective.

As far as the pressure is concerned, the lower the pressure the larger the spots will be, while the higher the pressure the smaller the spots. It should also be remembered that the distance that the gun is held from the surface of the substrate also dictates the effect of the spots, and varying this distance will give either a concentrated or dispersed effect.

Misting or Fog Effects (Sfumato)
(Also see Chapter 12)

This effect can be achieved by any standard spray gun equipment by a fine adjustment of the gun to produce a fine mist spray of concentrated colour compatible with the whole effect. For example, a light oak finish of a semi-gloss clear lacquer can be given a spray pattern of a green shading around the edges of the substrate to make a pleasing effect. However, if using a cellulose finish, then a pigmented cellulose must be used to give the fog effect which is over-sprayed when dry with a clear cellulose lacquer. The intensity of the colour fog of shading is controlled by the fluid needle adjustment and also by the position of the gun from the substrate. Multicolour fog effects can be over-sprayed with a clear top coat finish, either gloss or semi-gloss to trap or fix the whole shading effect within a film of lacquer.

Netting or Veining Effects

This is a decorative effect of random spray of lines of colour, thick or thin, as desired. The pigment colour should not be over thinned but used in such a way that it flows from the gun in thin lines at an air pressure of, say, 25 psi, and experimentation to obtain the correct viscosity depends on the operator. Cellulose is ideal for this effect due to its fast drying qualities. For example, a dark green background with a netting effect of yellow, may give you an idea what can be achieved.

Multi-Colour Effects

This is carried out by first using a basic colour background of, say, a yellow, then, before the coating can dry over spray using a blue colour, but not trying in any way to obliterate the yellow. You can then over spray these two colours with, say, an orange colour, but again not trying to obliterate either, so that a random three colour pattern can be achieved. When applied to a large flat area this gives a very pleasing effect. There is no end to the combinations that can be achieved by using colour applied by a spray gun, and they can be applied to walls, panels, furniture, canvas, doors, etc. The surface coatings can be oil paint, acrylic paint, pigmented cellulose, etc, depending on the effect required.

Waterfall Effect

This is achieved by first spraying a slow drying surface coating such as oil paint on to a flat substrate so that the viscosity of the surface coating allows the paint to run. Immediately follow this by over spraying a different colour which also runs into the first colour and so on. The skill here is knowing when to apply the second or third colour, or the whole effect may look like an unstirred paint tin. The basic technique is that the colour blending must be controlled and a very pleasing effect of flowing water can be obtained. When all the colours have been allowed to dry, a spray coating of a gloss varnish will add to the overall effect.

Marble Effects

Using a standard spray gun some very attractive simulated marble effects can be achieved. The pigmented colour can be applied by spraying through hessian or net, an old piece of twisted rope, or a muslin screen, or even using a large piece of steel wool, and can give wonderful effects. It is here that the skill of the operator coupled with an artistic skill is required.

Suede Effect

This is a recent development by Sonneborn & Rieck Ltd in the UK in the form of a spray-on finish that has the look and feel of suede. The product Jaxalac VT3 is formulated as a two pack polyurethane, using an aliphatic polyisocyanate as a curing agent, which provides it with a durable resistance to normal use, and is available in a range of standard colours. The product can be applied by normal conventional or automatic spray systems, and the coverage is 4 sq metres per litre which can be either force dried for 15–20 minutes at 70–80°C/158–176°F or air dried within 24 hours. The finish is particularly useful for office furniture and a variety of household finishings, giving a smooth suede effect.

These are just some of the ideas that can be achieved using a spray gun and colour and the results can be fascinating as well as fun to do. However, do remember that when the work is done the guns and other equipment must be thoroughly cleaned using the compatible thinners of the material sprayed.

Health and Safety Check List

See chapter on Health and Safety for complete safety precautions using spray equipment.

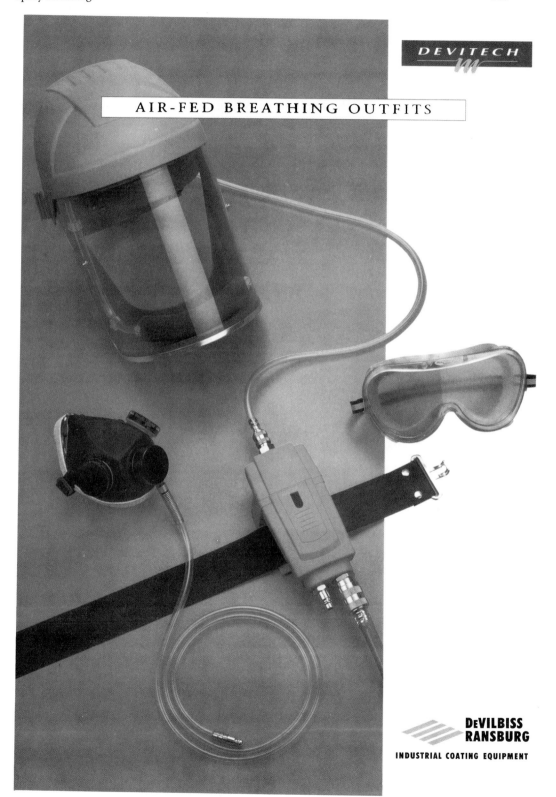

SPRAYING SAF[E]

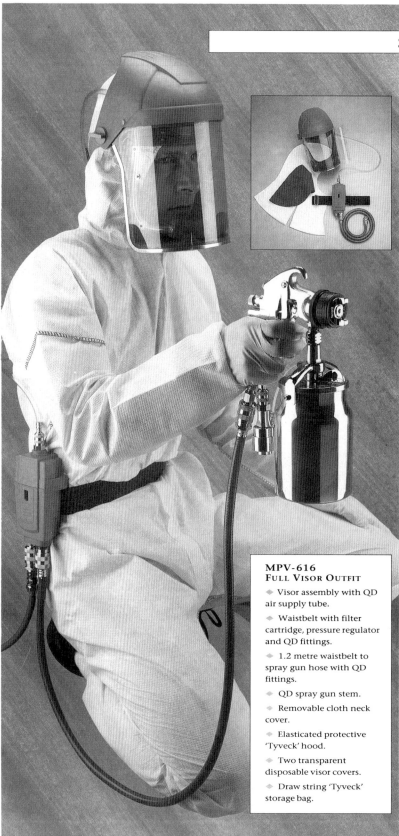

To maximise personal operator safety DeVilbiss Ransb[urg] have developed a range of air fe[d] breathing outfits to provide the most effective, practical and comfortable way to 'Spray Safe'.

Designed to meet all COS[HH] requirements both visor and ha[lf] mask outfits set new standards [of] protection and safety whilst allowing freedom of movemen[t] and causing minimum encumbrance to the wearer. Components are light, but stur[dy] and the range of adjustments ensure that total facial fitment is easily achieved.

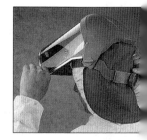

FULL FACE VISOR

◆ Compact, lightweight plast[ic] moulded construction provide[s] ultimate operator comfort.

◆ Larger visor offers all roun[d] vision and grade 2 impact resistance.

◆ High padded browguard an[d] detachable neck cover ensure complete head protection.

◆ Headband adjustment ratch[et] enables fast fumble free locati[on on] the head.

◆ Adjustable crown strip incorporates a convenient flip [up/] nod down facility.

◆ Carefully designed air disp[ersal] for increased comfort and qui[et] breathing.

◆ Comfortable foam seals maintain slightly positive air pressure and eliminate condensation.

MPV-616
FULL VISOR OUTFIT

◆ Visor assembly with QD air supply tube.

◆ Waistbelt with filter cartridge, pressure regulator and QD fittings.

◆ 1.2 metre waistbelt to spray gun hose with QD fittings.

◆ QD spray gun stem.

◆ Removable cloth neck cover.

◆ Elasticated protective 'Tyveck' hood.

◆ Two transparent disposable visor covers.

◆ Draw string 'Tyveck' storage bag.

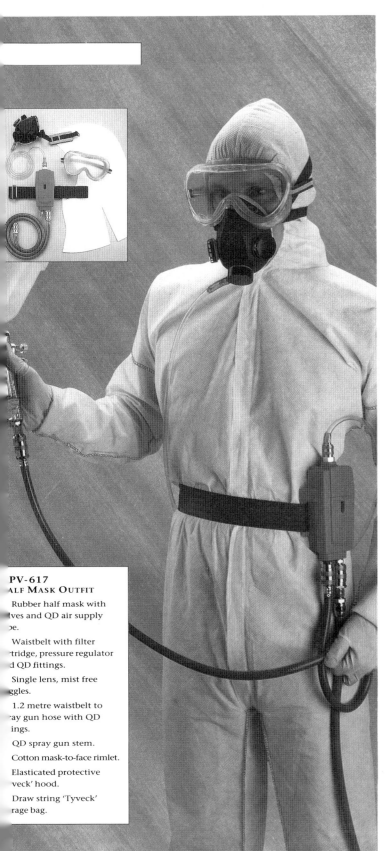

Half Mask

- Lightweight 'easy to wear' design allows excellent freedom of movement.
- Fully adjustable padded headstraps and pliable pinch at the nose provide effective facial sealing.
- Dual exhalation valves and air inlet diffuser eliminate obtrusive operating noise.
- Flexible air supply tube features a spring guard support for added safety.
- Single lens goggles with foam seal provide mist-free vision.

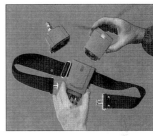

Multi–Purpose Waistbelt

- Routes incoming air for breathing and spraying from a single supply line.
- Provides reliable filtration and accurate pressure regulation of breathing air supply.
- Preset tamperproof regulator maintains constant air flow of 180 litres/min at operating pressures up to 100psi.
- Second stage high absorption carbon cartridge removes final traces of oil vapour and odour.
- Long life expectancy of up to 1000 hours before cartridges need replacing.
- Reliable colour change indicator provides visual confirmation of a safe breathing air supply.
- Convenient quick detach couplings assist off the job mobility.

Reproduced by kind permission of The DeVilbiss Company Ltd

PV-617
Half Mask Outfit

- Rubber half mask with valves and QD air supply pipe.
- Waistbelt with filter cartridge, pressure regulator and QD fittings.
- Single lens, mist free goggles.
- 1.2 metre waistbelt to spray gun hose with QD fittings.
- QD spray gun stem.
- Cotton mask-to-face rimlet.
- Elasticated protective 'Tyveck' hood.
- Draw string 'Tyveck' storage bag.

CHAPTER SIXTEEN

Spray Booths & Equipment

A spray booth is a semi-enclosed metal cubicle which has a system for extracting over-spray fumes, fog, solvent and acid gases. In areas where spraying is carried out, the surrounding air becomes contaminated with fumes of solid particles, gases, etc, and this is both undesirable and hazardous to the health of the operator. Therefore, spray booths are essential where continuous spraying is carried out.

Spray booths vary according to the type, shape and size of the objects to be sprayed, and they are made accordingly. A manufacturer making wheel barrows, for example, will require a small capacity spray booth, while a motor garage re-spraying cars will require quite a large system to accommodate the size.

It must be understood that an area where spraying is carried out is classed as a 'Spraying Area' and it should conform to certain conditions:

1. The spray booth must be in an area with fireproof doors and suitable exits so that operators can escape in the event of an outbreak of fire.

2. Fire fighting equipment should be on hand.

3. All electrical wiring, lights, switches, fans, etc, must be flash-proofed.

4. Operating switches for fans, etc, must be connected outside the spray booth.

5. An adequate clean air intake and extraction system should be provided.

6. It should be made possible to remove waste spray deposits when required or employ specialised companies to carry away the waste.

These precautions are basic common sense, but a good spray booth is generally installed by specialist manufacturers in the field, according to the specific requirements of the user's products.

There are, however, two main types of spray extraction system:

1. The dry-back spray booth,

2. The wet-back or water wash spray booth.

The Dry-Back Spray Booth

This is basically a metal cubicle with an extractor fan at the back. The spray booth has metal sides, the inside of which are coated with a skin of semi-adhesive latex film, which is sprayed on and can be peeled off when it has absorbed overspray deposits. This type of flexible skin is used in the film industry for outside locations, and no harm comes to the actual substrate after use. If latex film is not used then heavy layers of fire-hazardous, over-spray coatings would build up on the metal sides, creating a fire hazard. At the back of the spray booth is a belt-driven extractor fan through which air is drawn to remove fumes away from the article being sprayed. The extractor is fitted with disposable paper or fibre filters which collect the solids and waste material. An advantage of the dry-back system is that the filter beds can be replaced quite easily when new ones are required.

The Wet-Back or Water Wash Spray Booth

There are various types of water wash booths which basically work on a system of circulating water. The booth is built of metal and has a system where a continuous curtain of water flows down the back of the spray booth. The over-spray solids are collected in the water and settle as a sediment at the bottom of the collecting tank. Alkaline cleaning agents are added to the water in order to prevent the clogging of the jets and plumbing system. This system creates cleaner environmental working conditions for the operator. The periodic cleaning of the water plumbing system to remove the accumulated deposits does, however, take time, but it is still one of the most popular and safe systems used in the paint spraying industry.

Some water-wash spray booths have a system of removable residue collection panels on the back of the booth so that a minimum amount of time is spent on maintenance and cleaning.

Extraction Systems

The extraction of fumes, over-spray deposits, etc, must be carried out very carefully. If polyester or acid catalysed lacquers are being sprayed and a dry-back booth is being used, then the extraction may cause minute particles to drop onto the exterior extraction area, so great care is required when positioning the outlet to avoid, say, extraction dust falling on parked vehicles, etc.

When you choose a spray booth, therefore, you must bear in mind the actual process material being atomised, and you must make your selection with care as the wrong choice of booth can be very costly. No matter which type you choose, both are equally efficient in extracting spray fumes – it is simply a matter of choosing one that suits the situation in relation to surface coatings and extraction position.

When any extraction system is used in a workshop, air is removed at a rapid rate, and the temperature in the spray area can fall and cause problems on drying surface coatings. However, warm air can be fed into the spray area from an outside filtered source, thereby compensating for any heat loss.

Fig 63 Dry-Back Spray Booth (steel)
The long life dry filter is easily replaced, which can be a considerable advantage over wetwash booths which need some time to clean.
Photograph courtesy of P & J Dust Extraction Ltd

Fig 64 A dry filter or dry back spray booth
Photograph courtesy of Airflow Product Finishing Limited

Spray Booths

Fig 65   A waterwash spray booth
This system is an ideal method of extracting overspray making the working conditions pleasant for the operator. (NB No face masks are required using this system, subject to any special instructions from the lacquer manufacturers)
Photograph courtesy of Airflow Product Finishing Ltd.

It should go without saying that with an industrial workshop spray system, the materials used have a low flash point and are therefore a potential fire risk. It is absolutely crucial that care must be taken at all times by the operators when using these petroleum based derivatives.

Spray Tables

These can be placed in any type of spray booth, and are used for spraying small objects, such as chairs. The operator places the chair on a metal table, the height of which can be adjusted, and the top of the table can be revolved. Instead of the operator having to turn the gun to spray, the table top is turned, which means that the operator can keep the gun pointing into the back of the spray booth at all times.

On Site Manual Spray System

Not everyone has the advantage of a fixed spray booth, nor is it practical for spraying on-site substrates such as re-fitting bars, re-decorating, machine parts, structural objects, etc. In these circumstances, when a spray gun is a more appropriate tool than hand brushing to obtain a first

Fig 66 Standard Spray Table
Photograph courtesy of Gibbs Finishing Systems

class finish, life can be made easier for all concerned if operators of this equipment carry out the basic rules of the trade.

The choice of correct equipment is essential and the use of low pressure spray systems is an advantage. The goal with on-site work is to have as little over-spray as possible so as to avoid disruption and discomfort to others working nearby. Sometimes, this is not possible and this means that great care must be taken; face masks should be worn by the operators, fire fighting equipment must be on hand and in working order, and safety notices, such as 'no smoking', 'no naked lights', 'spray painting in progress', etc, should be displayed. The use of water-borne surface extractor fans and filters for over-spray collection are also beneficial under these circumstances. It is up to the operator to make as little nuisance as possible for himself and to others.

Machine Spray Systems

The Curtain Coating Machine
This is one of the most popular machine systems for applying finishes to flat substrates of wood, metal, plastics, etc. It is an industrial method designed to apply vast quantities of surface coating material to an object with a speed that would be impossible to match by normal conventional hand spray methods. It is ideal for finishing such items as panels, doors, wardrobes, indeed any flat object that is produced which requires fast, quality surface coating.

The 'curtain coating machine', as its name implies, provides a *curtain* of surface coating through

Spray Booths

which the substrate is propelled by a conveyor bed; to give a somewhat common comparison, it is like passing through a waterfall.

The basic construction is a conveyor bed which is split in two parts, and the trough holding the surface coating (lacquer, etc) is suspended in the middle of these two beds. As the object to be finished is fed on to the conveyor belt, it passes throiugh the curtain of, say, lacquer. The timing of the conveyor belt is coordinated with the viscosity of the lacquer and adjusted accordingly.

Outlined below are the type of instructions for a job that would be given to an operator for a conventional polyester system using a curtain coater:

Substrate: veneered covered chipboard panels for use as sides of wardrobes.
Total Number of Panels to be Finished: 150
Process: Surface coating: Pigmented Polyester

1. Apply by curtain coater 10 gms per sq ft. (100 gms per sq metre) of activated ground coat.

2. Leave to dry for 1–2 hours.

3. Apply by curtain coater 40–50 gms sq ft (400–500 gms per sq metre) of pigmented polyester.

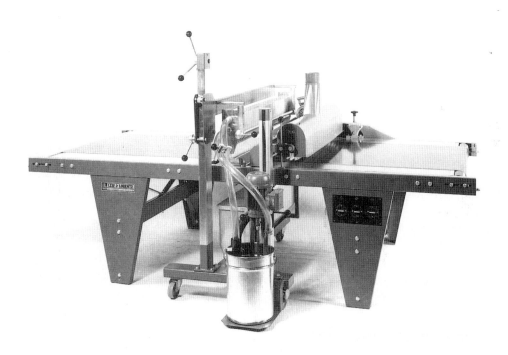

Fig 67 Curtain Coater Machine
Constructed for use in both large and small industries, rendering an efficient and quick surface treatment of workpieces such as wood, metal, glass and plastic. The machine is 100% efficient and requires minimum cleaning.
Photograph courtesy of Leif & Lorentz, Denmark and supplied by Gibbs Finishing Systems of London.

Leave from curtain coater for a minimum of six hours before sanding and full gloss burnishing. (Force drying would reduce this time considerably.)

Excluding drying time, the whole job could be finished by two operators in approximately one hour, with one feeding the panels on to the bed, and the other removing and stacking them as they came off the end completed. The speed that the panels run through the curtain coater is quite staggering, and the two operators must work in harmony if the finished articles are to end up properly stacked.

Trolley Racks

These are metal frames which have rods at right angles that act as shelves, the whole unit being on wheels for mobility. These racks are ideal when used in conjunction with a curtain coating system; the panels coming off the conveyor belt must be stacked onto racks as quickly as possible and moved away as soon as they are full for curing. When not in use, the trolley racks take up a minimum of space as they fit into one another for storage. They can be used with any form of spray system and are particularly useful where the finished substrates need to be stored carefully for drying immediately after being sprayed.

Fig 68   Trolley Racks
Provides maximum capacity for drying panels, with
minimum storage space when not in use.
Photograph courtesy of Gibbs Finishing Systems

Roller Coating Surface Coating Machine

This is another system for applying surface coatings to flat substrates. It consists of two rollers made of a hard plastic material which is not affected by chemical substances. The rollers are mounted vertically, one over the other, and are rotated mechanically. The gap between the two rollers can be regulated for varying thickness of the substrate. Above the top of the roller is a reservoir in the form of a trough which holds the surface coating fluid, which drops on to the top roller so that the substrate passing through the rollers is coated with a deposit of surface coating.

This system is frequently used to apply sanding sealers because the roller pressure forces the coating into the grain of the wood, which is not possible with the curtain coater system. However, the substrates must be flat, for if there is any fault on the surface the roller will pass over it so no surface coating will be deposited; in other words, the roller deposits the surface coating only where it touches.

An extra roller can be fitted as an addition to the system which rotates in the opposite direction. It is used immediately after the substrate has been coated and its function is to remove unwanted surplus material. Another way of achieving this is to use a blade which scrapes off the surplus coating material. The reverse roller system is used mainly on quantity coating runs where pigmented priming surface coatings are required.

Automatic Spray Machines
If the substrates are of varying shapes, such as mouldings, then an auto spray machine is used. This machine has a conveyor bed which passes the substrate through a unit housing two or three spray guns. Unwanted surface coatings can be collected and re-used. The spray system can be constructed to spray on all three sides of a substrate automatically. The spray guns are controlled to spray the varying shapes by photocells and timers. The speed of the conveyor belt can vary from 30–150 m/min.

Programmable Robot Spray Machines
These machines are robots that reproduce the movements of spraying carried out by hand. They record the different movements that would be done if a human operator was using a spray gun. Once the robots have been set up and programmed, they can be used to spray such items as cars, machinery, production line components, such as domestic appliances, and many other mass produced objects that would otherwise be carried out by human operators using manual spray guns, but in a faster, more constant and efficient way. The system consists of a control cabinet, a manipulator and a hydraulic section. One company in the UK specialising in these machines is Kremlin who produce various systems for industry. They can be easily installed in existing production lines without modification, and are complete spray systems in themselves, although extraction of fumes, etc, is another matter. The great advantage of robots is that they can out-perform any manual spray volume work. Furthermore, they can supply a constant quality spraying performance. In the US many companies manufacture similar systems.

Robot spray programmed systems are particularly useful in areas where the working conditions for manual spray operators would be somewhat unpleasant. They are widely used in mass finishing industrial production lines.

CHAPTER SEVENTEEN

Health and Safety

During the last three decades our legislators have placed increasing emphasis on protecting the health of operators whilst using chemicals, oils, solvents and other potentially harmful substances. Trade training colleges have also been instrumental in teaching basic health and safety factors to warn students of the hazards they will encounter in their chosen trades whilst using products and equipment, and carrying out processes, that can produce dust, vapours and odours, as well as being potential fire hazards. When you look in detail at the many regulations and itemised warnings, the instructions are basically straightforward and simple common sense. Unfortunately, we humans have a habit of ignoring instructions and advice, which can be disastrous to both ourselves and others.

In the UK in 1974, due in part to new technology in the manufacture of solvent materials, an effort was made to introduce in one Act, as far as possible, a coherent set of regulations to cover the use of these materials, making manufacturers and suppliers responsible for correct labelling for the use of their products. The Health and Safety Act was introduced in 1974.

The first of the regulations relating to dangerous substances was introduced way back in 1928, and was subsequently added to as follows:

1. Petroleum (Consolidation) Act 1928 and subsequent orders.

2. Highly Flammable Liquids and Liquified Petroleum Gases Regulations 1972.

3. The Classification Packaging and Labelling of Dangerous Substances Regulations 1984.

4. Road Traffic Carriage of Dangerous Substances in Packages, etc – Regulations 1986.

As with most Acts, they were, in time, found to be wanting and further regulations relating to operatives at their place of work were introduced in 1989, called the C.O.S.H.H. (Control of Substances Hazardous to Health) 1989. These are the most extensive range of regulations ever introduced for workers in a variety of occupations, trades and professions. They bind together all the important regulations of past years and improve them to form a comprehensive set of regulations relating to man, machines, products, processes, equipment, transport, etc. This is, of course, only a very brief description of the regulations; the complete text is very detailed and requires, at times, a legal mind to understand the complexities.

In the USA the Environmental Protection Agency (EPA) develop guidelines for the State Environmental Protection agencies which in some ways are more complex than those of the UK, and more severe. A 'VOC' mark (Volatile Organic Compounds), is always marked on products which

Health and Safety

can produce ground level ozone, which forms smog, and products such as varnishes, lacquers, etc, must be low odour (VOC) compliant, according to the regulations.

A Guide to COSHH

Many wood finishers, woodworkers, builders, joiners, carpenters, construction workers and D.I.Y. enthusiasts have, until recently, suffered from the most deadly of all infections – that of indifference to health hazards, but an increase in public awareness of the health hazards associated with dust, fumes and solvents in recent years has led the UK Government to improve the safety regulations for the use of these products in both industrial and commercial processes. The new Health and Safety Executive regulations, C.O.S.H.H. came into force on 1st October 1989. The Act now places the responsibility for safety on employers and on the self employed or anyone working additionally on the premises of the employers.

The regulations cover over 40,000 substances which are used in manufacturing or construction processes. They lay down essential requirements and are a straightforward approach to the control of hazardous substances. Employers are required to:

1. Assess risks in actual working processes and determine what precautions are required.

2. Introduce measures to eliminate possible risks.

3. Ensure that the procedures are maintained and that any equipment used is properly maintained.

4. Instruct and train all employees of risks and the precautions to take. Failure to comply with C.O.S.H.H., in addition to exposing employees to risks, constitutes an offence and is subject to penalties under the legislation.

The enforcing authorities are the local Health and Safety Inspectorate, or the local environmental health departments of local councils, District or Borough.

A C.O.S.H.H. assessment must be made if hazardous substances are in use. This involves gathering together information on the products being used, as well as on the work practices associated with their use, and identifying any risk to the health of employees so that action can be taken to eliminate or control the problem. Fact sheets or information must be obtained from the individual manufacturer of the products or equipment being used in order to carry out an assessment. When all the relevant information has been gathered, the assessment is recorded so that it can be amended if further equipment of materials are introduced which differ from the originals. It is compulsory for all employers, as well as the self employed, to monitor hazards such as dust, vapours, etc, and make sure that they are within the maximum exposure limits laid down for the materials being used.

For those working in the wood finishing trade, there are a number of practical considerations to be taken into account when carrying out a C.O.S.H.H. assessment, such as how a particular product should be applied (brush, spray or roller), and how fumes, dust and vapours produced in stripping and preparing a substrate will be removed. All these points must be taken into account when completing an assessment.

Further information

* The complete Control of Substances Hazardous to Health (C.O.S.H.H. Approved Code of practice No 29 ISBN 011 885468 2 HMSO.
* C.O.S.H.H. Assessments: a guide to assessments, etc, ISBN 011 8854704 HMSO.
* Respiratory Protective Equipment: A Practical Guide to Users ISBN 011 8855220 HMSO.

(The above can be obtained from booksellers or the publishers.)

Glossary of Terms

The following is a glossary of abbreviations and terms which may be encountered when carrying out a C.O.S.H.H. assessment:

Abbreviations

Assessment:

P	Practically non-harmful – no further action required
I	Insignificant exposure – no further action required
W	Assessment carried out by another Trade
M	Medical surveillance required
RPE	Respiratory Protective Equipment required
PPE	Personal Protective Equipment required (exc. RPE)
LEV	Local exhaust ventilation equipment required

Occupational Exposure Limits (OELS):

MEL	(Maximum Exposure Limit)
OES	(Occupational Exposure Standard)
SUP	(Suppliers standard)
$>=$	more than
$<=$	less than
SK =	material absorbed through the skin (for example, methanol, butanol, xylene, toluene, isopropanol, etc)
$Mg.m^{-3}$	Milligrams per cubic metre = weight of solvent in the atmosphere in milligrams per cubic metre of air
PPM	Parts per million = parts of vapour/gas to one million parts of contaminated air by volume

CPL (Classification)

Health and Safety

> Very Toxic
> Toxic
> Harmful
> Corrosive
> Irritant
> Highly Flammable
> Extremely Flammable
> Oxidising

It is important to note that in most chapters of this book I have given a health and safety check list which deals specifically with the processes involved and which should be followed as far as possible by the operator. They are of my own design and the precautions suggested are based upon common sense; they do not form a part of any official assessment but are simply a quick check guide-line. It is up to the operator, therefore, to make his or her own assessment, taking into account the individual circumstances prevailing.

A guide to the safe handling of chemically based products used by polishers and wood finishers in general

All the materials used by wood finishers are potentially dangerous; a tin of chemical strippers, for example, left in direct sunlight could easily become a fire hazard. What follows is a guide to preventing potential dangers from turning into accidents:

1. Ingestion of chemicals should always be avoided. No food or drink should be brought to, stored, prepared or consumed in the area where chemicals are used.

2. No smoking in the work area or where chemicals are stored.

3. Washing facilities, such as hot water and hand drying facilities, should be available at all times for all working employees.

4. Facilities should be available for workers to change into and out of protective clothing.

5. Inhalation of vapours, chemicals, dust, etc, should be avoided. Ventilate all areas and use extraction systems. Adequate respirators should be worn.

6. All containers and products, etc, should be clearly and accurately marked.

7. All flammable products should be stored in a cool, dry place which is well ventilated and separate from any work area. With regard to storage of these products, structural requirements conforming to regulations should be correctly observed.

8. After use, all flammable products should be kept in fireproof bins inside the workshop area and disposed of daily.

9. A first aid box should be available at all times and a record kept of any minor accidents. Local doctors and emergency phone numbers should be clearly displayed.

10. Do not allow paint, bleaches, oils, solvents, etc, to enter main drains or watercourses.

11. All waste materials must be treated as a fire hazard and disposed of in accordance with the general requirements of the Control of Pollution Act 1974 and other regulations. (US and other countries will have their own specific regulations covering these requirements.) Burning waste deposits from daily workshop use can produce gases and hazardous vapours if not incinerated in a proper manner.

12. Wet flatting, instead of dry, should be used whenever possible in the preparation of wood and surface finishes to avoid the creation of dust.

13. Fire prevention: Fire is a hazard associated with either the storage or handling of chemical products, together with the processes involved in using them. The measure of flammability is by 'flash point', which is the temperature at which materials can be ignited or set on fire by applying a flame under certain conditions. In addition to the obvious hazard of the flames, flammable vapours and toxic fumes are also given off by many materials. Care should be taken with unprotected electrical equipment, especially where there is an accumulation of over-spray deposits, as well as with contaminated rags and waste material which could result in a possible electrostatic ignition. Keeping the workshop as clean as possible is, therefore, of paramount importance.

14. Fire fighting equipment should be kept in working order and employees given regular instruction on its use. Foam or dry extinguishers, together with buckets of sand, should always be on hand.

15. Good natural ventilation is advisable at all times to prevent a build up of concentrations of vapours, even in areas where chemicals are stored.

16. To prevent dust inhalation, wear face masks, goggles or face shields at all times.

17. Use protective plastic gloves for all purposes, such as staining, etc.

18. Respiratory equipment: Where spray equipment is being used, extraction spray booths or respiratory protective equipment, which include compressed air supplies, should be selected, used and maintained in accordance with C.O.S.H.H., O.S.H.A., and other local regulations.

19. Personal Hygiene: Anyone using wood finishing products should always wash hands with soap and clean water before consuming food and drink. After washing hands, it is advisable to use a hand cream to replace natural oils in the skin which are removed when handling solvents, and make the skin very dry. It is also advisable to use barrier creams before starting work. Smoking should be banned in areas where solvent based paints, varnishes, etc, are being used, not only due to the risk of fire, but also that of inhaling burnt solvent fumes and by-products. Overalls should be removed before meal breaks and laundered regularly. Meal breaks should always be taken away from the work area.

20. Skin complaints – Dermatitis

There are two types of dermatitis – irritant and allergic.

(a) *Direct Irritant Contact Dermatitis*

Definition: dermatitis caused by exposure to a substance which has a damaging effect on the normal barrier function of the epidermis (Mackie, 1986).

Irritant dermatitis may be an acute reaction to one single exposure to a strong irritant, such as an acid. Onset is rapid and the lesions appear on the skin where the skin was in contact with the irritant chemical.

A second variety of irritant dermatitis results from cumulative exposure to a mild irritant. A common example is detergent. Mild examples of cumulative irritant dermatitis are extremely common among people regularly exposed to detergent/paint at work or at home.

Presentation

acute: After exposure to a strong irritant the affected skin becomes reddish and blisters can develop. This usually happens within 6–12 hours following contact. If there is no further contact with the irritant then recovery is usually rapid. Identification of the irritant substance is usually made by the individual concerned.

chronic: This usually presents as dry, fissured skin which is susceptible to secondary infection.

Treatment

Identification of the irritant and future avoidance is necessary. Calamine lotion may be helpful. Topical steroid creams may be prescribed.

Prevention

Prevention of irritant dermatitis is simple. Wash hands well with soap and water after contact. Wear rubber gloves with cotton gloves underneath to absorb perspiration, when in future contact with the irritant. Barrier hand creams may be helpful. Face masks may need to be worn if abrasive dust is the cause of the problem.

(b) *Allergic Contact Dermatitis*

Definition: dermatitis caused by *prior* exposure to an allergen leading to sensitization.

Incidence is quite common It affects 1–2% of the population, e.g. workers exposed to materials such as chromates in the building industry or dyes used in the leather industry have a high incidence of this form of dermatitis. Other common allergens are rubber, oil, wool, alcohol, solvents and epoxy resin.

The commonest agents have been identified and are used in patch testing for investigating sufferers of allergic dermatitis.

Presentation

It may develop after many trouble-free years of contact with the allergen. It may be hard to convince someone, even after positive patch testing, that the cause of their skin problem is something that they have used for many years. The skin is commonly red and itchy and continued exposure will lead to dryness and scaling. Hand and forearms are commonly affected. Face may be affected.

In theory, once identified the individual should avoid contact with the substance. It is often difficult to avoid allergens such as nickel, for example, as it is found in coins and cutlery used in every day life. Re-employment may be necessary, but difficult of course. Factory or Industrial Medical Officers may be of assistance.

Treatment

Patch testing to identify allergen. Avoid contact with allergen. Counselling. Short term use of steroid creams.

References

(1) Mackie Rona. Clinical Dermatology
An Illustrated Textbook 1986 published by Oxford University Press
(2) Hunter, Savin, Dahl
Clinical Dermatology published by Blackwell Scientific Publications 1989

21. First Aid:

Foreign bodies in eye: medical aid is needed (i.e. for wood, metal splinters).

Chemicals in eye: wash eye with copious amounts of cold water. Very painful condition. Cover the eye with a pad and seek medical advice immediately.

Inhalation of fumes: open windows and get the individual outside into fresh air. Depending on the severity of the condition, medical advice may be needed.

Skin burns from chemicals: wash with copious amounts of cold water and cover with non-adherent dressing. Seek medical advice.

Lacerations: Venous, i.e. oozing with blood. Apply clean, non- adherent dressing over the site and apply a firm bandage. Raise the limb to reduce the bleeding. Arterial, i.e. pumping blood from injury. Apply digital pressure over a dressing pad over the site and keep the pressure. Do not use a tourniquet. Get medical help *immediately.*

Ingestion of poison/chemicals: Do not make the individual vomit as the substance may be caustic. Give the individual lots of water to drink and seek advice from a hospital Casualty Department. Take the bottle or write down the name of the chemical ingested when accompanying the individual to Casualty.

Electrocution: Turn the power OFF before touching the affected individual. If this is not possible then use a wooden broom handle, for example, to remove the individual from the power source. Then turn the power off. Check if the individual is breathing and has a pulse. Talk to the person, if he is conscious and alert, then leave him/her lying down and cover the electrical burn with a dressing pad and seek medical advice.

If the person is breathing but is unresponsive, then place him/her in the recovery position, i.e. on their side. Cover him/her with a blanket, cover the electrical burn with a dressing pad and call for urgent medical help.

If the person is not breathing and no pulse can be felt, then call for urgent medical help and commence immediate resuscitation, i.e. mouth to mouth and heart massage. Keep this going until the person recovers and starts breathing again or until medical help arrives. This resuscitation may save a person's life.

Suggested contents for a first aid box in a small/moderate sized workshop:

cotton wool

2 crepe bandages 75mm/3" width

Health and Safety

scissors
safety pins
antiseptic wipes
antiseptic cream
pack of mixed size plasters/bandaids
roll of zinc oxide 25mm/1" tape
2 gauze bandages 50–75mm/2–3" width
2 non adherent dressing pads 100mm/4" square
1 triangular bandage
eye bath
2 eye pads
pair of fine forceps/tweezers
1 bottle of Paracetamol or Asprin tablets
First aid guidance information
Illustrated brochure/leaflet of mouth to mouth resuscitation techniques
Record book of accidents/treatments

Fire Fighting Equipment

Fire Extinguishers

Quick check reference

Anyone working with low or high flashpoint flammable materials must have a fire extinguisher available for immediate use. If a fire starts, seconds are vital and it is no use thinking 'where is that fire extinguisher', it must be available where anyone can see it, and be in working order. When fire occurs there is no time to read the instructions on the fire fighting equipment, you must know how to use it without a second thought. This is where staff training is essential in the use of any fire fighting equipment.

There are, however, four basic types of fire classification:

(a) FIRES. Where wood, papers, textiles, etc are burning.
(b) FIRES. Where flammable liquids, grease and oils, etc, are burning.
(c) FIRES. Where gaseous fires are burning.
(d) FIRES. Where mainly electrical equipment fires are burning.

The correct type of fire extinguisher to be used are as follows: (The actual fire extinguishers are painted in colours to denote the type).
CREAM to use on (b) fires filled with AFF (Aqueous Film Forming Foam)
RED to use on class (a) Fire. These are water filled type extinguishers.
BLUE to use on class (b) and (d) Fire. These are dry powder filled under pressure.
BLACK to use on class (c) Fire. These are filled with (CO_2) vapour forming gas. Carbon dioxide use on (d) live electrical equipment.

(The dry powder type extinguishers are most popular, and can be obtained in various sizes, to suit the situation). **Note:** Colour identification may vary in different countries.

The Sand Bucket

These are always handy for a very small type fire where speed is essential; the sand must always be dry.

Fire Blankets

These are fire resistant and ideal for small area fires such as burning oils, etc.

(Speed in the application of a fire fighting substance is of the paramount importance when dealing with an unexpected fire, the equipment must be kept clean, checked regularly and must be in an accessible position.)

The above details are meant as a guideline only and where in doubt a specialist company should always be contacted who can advise the correct type of fire fighting equipment suited best to your situation, or advice from your local fire station can always be useful.

Health and Safety Precautions when using a Spray System
(To be used mainly in conjunction with Chapter 15 'Spray Finishing')

These are what I call the ten commandments for anyone using a spray system whether they are classified as a large concern or a small workshop unit, or indeed the individual or single craft person. The principles of health and safety precautions are applicable and must be identified and strictly observed.

1. No smoking either in or anywhere near the spray area.

2. Make provision for a suitable air over-spray extraction either by a spray booth or other means, and also for clean air change, and clean intake air.

3. Wear suitable face masks, goggles, visors, air-fed breathing outfits, etc, suitable for the safe use of the materials being sprayed. This is particularly important when spraying with polyurethane lacquers.

4. Make sure that the extraction of waste fumes does not contaminate any other areas where people are working, thus avoiding this nuisance.

5. Read the instructions and plan the use of the materials to be sprayed.

6. Do not drink or consume food whilst in the spray area.

Health and Safety 205

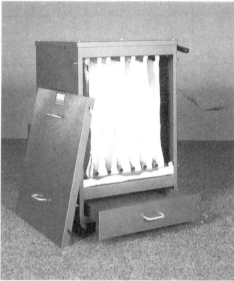

Fig 69 Extractor for Fine Dust Particles
This robustly constructed Extraction Unit is designed to efficiently extract the waste from fine dust producing machinery.
Photograph courtesy of P & J Dust Extraction Ltd.

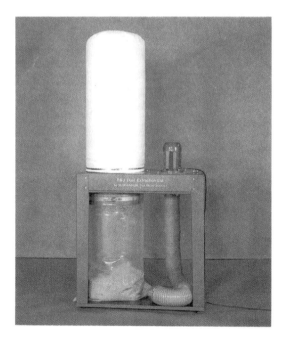

Fig 70 Dust Extractor
Designed to provide low-cost collection and disposal facilities for small workshops.
Photograph courtesy of P & J Dust Extraction Ltd

7. When spraying the new water-borne water based lacquers, face masks should be worn.

8. In the spray area make sure that all electrical equipment, lights, fans, switches, etc, are correct for use in high risk areas to eliminate sparking. Also make sure that any compressor is suitably screened away from the spray area. When using low flashpoint materials operators should not wear metal heels on their shoes so as to avoid sparking on concrete floors.

9. Have suitable fire fighting equipment at hand, also buckets of dry sand (not water).

10. Keep visitors, pets and children away from the spray area; persons who suffer from any form of asthma should never use or be employed in or near a spray system area.

CHAPTER EIGHTEEN

Questions and Answers

Questions and Answers to Wood Finishing Problems

For the last 30 years I have been answering questions sent to me from readers of magazines, students, builders, polishers, industrial concerns, members of the general public, local authorities, architects, kitchen fitters, parsons – the list is endless. Sometimes, the information I was sent was less than I would have liked and often the most vital facts left out. Accordingly, some answers may reflect the difficulty in providing solutions to problems where insufficient details have been given.

Sometimes, my replies had a humourous content; some were short and to the point, whilst others were more complete. In selecting the letters for this chapter, I have chosen those which I think will be of interest to the reader, and I hope they will be found to be a relaxing change from all the facts and figures in the book as a whole.

Question

I am making a fruit bowl in ash and I want it to be stained black. Would you please suggest a non-toxic stain which will not obscure the grain and will give a suitable dull finish?

Answer

Ash! What a wonderful wood to make a fruit bowl out of, with its straight grain and coarse texture.

I am sure that you will see to it that the bowl is in a perfectly sanded and prepared state before you embark on the finishing. What I suggest you do next is to stain your bowl with a black aniline water stain made up with warm rain water (which is softer than tap water), plus a drop or two of household ammonia to help penetration. Always test the stain on a spare piece of wood first to make sure that you have the desired shade.

After staining the bowl, allow it to dry out. Do not at this stage sand down the resultant finish as you will have found that the grain has been slightly raised. In your letter you do not say if you require an open grain or a filled grain effect, but as the bowl is for fruit, I would suggest that the grain is filled for hygienic reasons. With a brush apply one coat of a pre-catalysed sanding sealer, which is quite thick, and when hard flat down using 240 grade garnet abrasive paper. If required, apply a further coat so that the grain is completely filled. Next, apply one good brush application of a semi-gloss brush-on pre-catalysed clear lacquer. When dry, flat down using 320 and 600 grade wet and dry abrasive papers, using a little rain water as a lubricant, then pull over using a rubber and strong pullover fluid and leave to dry. The desired effect of 'shine' can be obtained by using 0000 steel wool.

You will end up with a finish that will withstand the effects of fruit and allow the bowl to be washed in water without damaging the surface finish.

Question

I am repairing a pair of gates for a church, which are made of oak, and must be at least 400 years old. Whilst I am confident that I can repair the gates, I do not know how to finish the wood. As they stand now, they are very grimy due to the passage of time and the weather, and have never been coated with any preservative. What is the best way to blend the new wood with the old and to preserve the wood with a finish?

Answer

You say that you are confident in carrying out repairs to these gates, which is an admirable attitude to take. However, with all due respect, there are many problems which an amateur could encounter, so I hope you will find the following comments helpful.

First of all, you must wash down the old part of the gates with a solution of warm water and mild disinfectant (follow the directions for making up the solution on the tin as for cleaning greenhouses, etc). Make sure that the mixture is scrubbed well into the fibres of the old oak, and allow to dry out thoroughly. This process will kill off any insects and fungi.

After cleaning the timber, you will be able to see what repairs are needed and can mend the gates using good quality, well-seasoned oak. Do make sure that your repairs are compatible with the original workmanship – if the mortice and tenon joints were dowelled, for example, then 'dowel' your new joints. It really goes without saying that no glues should be used on this project, so your workmanship must be of the highest order. A further point to remember is that all screws, nails, bolts, etc, must be of the non-ferrous type; oak contains a large amount of tannic acid, which reacts with metals such as iron or steel and quickly rusts, leaving a black stain around the site of the metal. In view of this, choose copper, brass, stainless steel, etc, for your ironmongery.

Turning to the question of preservation; you say that the gates have lasted 400 years without any finishing, so do likewise – leave them alone. My only reservation is that until the new wood matures, which may take some time, it will stand out like a poppy in a wheat field, so I would suggest that you simply stain the new oak using a chemical stain which will be everlasting. Make up a solution of warm water and sulphate of iron, known in the trade as Green Copperas, adding the powder in equal parts by volume of water. Apply the stain to the new oak and leave it to dry out. It is always a good idea to try the effect on an offcut first so that you can adjust the mixture to the colour you require. When dry, the new oak will have taken on the effect of old, weathered, silvery grey oak,

Questions and Answers

which is permanent. This will age, in time, and blend in to the older parts of the gates. The beauty of chemical stains, unlike other stains, is that they penetrate deep into the fibres of the wood and react with the chemical contents of the wood and become compatible with it. Leave the restored gates to weather, and who knows – they could last for another 400 years.

Question

I have a small carved table of oriental origin which has a finish in matt black. Over the years it has become chipped and shabby, and I want to re-finish it and restore it to its original state. How do I remove the original finish (which, incidentally, I cannot identify), what do I use to re-coat it, and where can I obtain the materials from?

Answer

Re-finishing wood carving requires patience and caution in the operation.

You state that the surface of your table is chipped and worn and that you do not know the original surface coating. This eliminates any kind of touching up of the damaged surface. You also state that the table could be of oriental origin which makes me believe that it is a shellac-based surface coating as shellac is used extensively as a wood finish in the Far East regions of the world.

This old surface should be stripped down to bare wood, for which I would recommend a standard, thin chemical stripper, instead of the thicker water-washable type. Any use of water could cause damage, such as splitting to the wood carving, and should be avoided. Other materials required are waste cotton wool, or absorbent paper such as kitchen paper towels, cellulose thinners, white spirit or turpentine substitute, a paint brush or grass brush, a quirk stick (a small pointed stick for getting into cavities), an old tooth brush or soft brass wire brush.

Proceed as follows:

1. Apply chemical strippers liberally by brushing well into the carving.
2. Leave on the surface for at least 15 minutes.
3. Wipe off the soft 'gunge' using the cotton wool or absorbent paper.
4. Use the cellulose thinners to wash out the chemical strippers from the carving cavities with the help of the paint brush.
5. Use the wire brush, quirk stick and toothbrush to complete the cleaning operation.
6. Wash down with white spirit or turpentine substitute.
7. Allow to dry out naturally.

On no account use steel wool on this operation as the thin wires can become trapped within the carving, and have plenty of ventilation when using any of the chemicals mentioned. Wear rubber or

PVC gloves and observe normal fire precautions when using these low flashpoint fluids – and remember, no smoking and no naked flames anywhere near the work area.

To re-finish the table, I would recommend a cellulose black matt aerosol, which is obtainable from any car accessory or D.I.Y. store and proceed as follows:

1. When the substrate carving is bone dry, apply one thin spray application of the cellulose black matt and leave to dry for approximately 15 minutes.

2. Apply further thin coats at 15 minute intervals until you are satisfied with the appearance. Do not be tempted to shorten the process by applying a thicker coating as this could result in pools of cellulose collecting in the cavities causing uneven drying out and consequent cracking of the cellulose coating.

3. Allow the whole job to dry out naturally and keep out of direct sunlight or blistering could occur.

Do not use the table for at least two to three days after the last spray coating.

Question

I have a 1950's wardrobe made in oak veneer, and I would like to renovate and re-finish it with a more modern texture. Ideally, I would like to achieve a light grey stain which would leave the grain showing to its best advantage. I wonder if you could tell me how to achieve this finish and, in particular, how the base colour of the oak is prevented from filling the grain.

Answer

There are various methods of achieving the finish that you require which is so popular with commercial furniture manufacturers of today. One such method which I recommend you to try is detailed below.

1. First, remove all old surface coating film from your wardrobe by using a chemical stripper which is better to use in this case, due to the veneers, as no water is used to neutralize. After removing all traces of old surface coating film you must neutralize the action of the stripper with white spirit and allow to dry out. It goes without saying that you follow the manufacturer's instructions on the tin to the letter. (Any good brand of chemical stripper can be used.)

2. When dry, the substrate grain is now scrubbed with a good quality soft wire brush. The object of this is to remove all traces of soft fibre and waste material from the grain cavities. The best type of wire brush to use is a brass one because you must take care not to rub through the veneers. Never scrub across the grain but work in the direction of it. Do not on any account sand down or you will simply fill the grain up with sanding dust.

3. Brush down the substrate, which is now ready for re-finishing using a modern surface coating. Apply by spray gun or aerosol can a light coating of a cellulose white wood primer and allow to dry out for at least two hours. Do not flood coat this primer or it will flow into and fill the grain cavities.

4. Apply by spray gun or aerosol can a light spray coating of grey pigmented cellulose satin surface coating and again do not flood coat – it is easy to do when spray coating. Spray just enough to cover the basic oak leaving the grain unfilled or open. Allow to dry out for at least 12 hours in a well ventilated room at, say, 65°F.

It should be pointed out that these basic pigmented cellulose finishes require thinning with cellulose thinners to obtain the correct viscosity – approximately 10% is added to the basic colours. This does not apply, of course, if using an aerosol.

Remember that these materials have a low flash point and care must be taken with ventilation, extraction of fumes, etc. Under no circumstances should you smoke in the vicinity of these products.

If you follow the steps outlined above, you should achieve the finish that you requested – a grey basic background with the natural oak grain showing to its best advantage.

Question

I have been given a pine dresser which has been in a caustic (lye) stripping bath, but I have noticed that the pine appears to have taken in fine deposits of what looks like salt. There is no finish at the moment on the pine but I am wondering what I should do before applying a varnish finish.

Answer

Your problem can be solved by following the procedure:

1. Wash the affected 'salted' areas of the pine dresser using a mixture of warm water and a little detergent and scrub into these areas liberally.

2. While the wood is wet, apply a solution of mild acetic acid or diluted white vinegar and leave this solution to lie on the wood for at least 1 hour.

3. Wash the treated surfaces thoroughly with clean water. If possible use running water from a hose pipe and a scrubbing brush.

4. Allow to dry out naturally for a day or so.

5. Bring the dresser into a warm dry atmosphere at a temperature of not less than 65°F and leave to dry out for a further 48 hours.

6. Apply heat from a hot air blower if required until you are sure that the wood is bone dry (this is important).

7. Sand down using 100 grade garnet papers and finish off with 240 grade papers.

8. Apply two coats of dewaxed shellac sanding sealer to the whole dresser and leave to dry out for 48 hours.

9. Re-sand lightly using 320 grade Lubrisil abrasive papers and dust off.

10. You can now apply any good quality clear gloss or satin internal varnish – preferably *not* a polyurethane type. Two coatings should be sufficient. Leave to dry out for 3–4 days in a temperature not less than 65°F. You should now have no further trouble.

Question

I have made a cabinet in walnut veneer and I want to finish it in full gloss black varnish lacquer, similar to the Eastern black lacquer finishes, such as japanning. How do I make up such a finish?

Answer

Before I answer the question, a few words about japanning in connection with wood finishing.

Japanning is a process originating from China some 2000 BC but developed by the Japanese who became masters of the techniques. The method involved covering a substrate with 15 to 20 layers of 'lacquer'; each coating film was flatted and burnished before further coats of lacquer were applied. The original lacquer was shellac varnish over black pigments, or a mixture. When the substrate was dry and burnished it was decorated with gold or gilt. English lacquer, which was made up of copal resins and drying oils such as linseed and tung oil, was developed later. It is interesting to note that the technique was developed in this country through a workshop book entitled 'A Treatise of Japanning' by John Stalker and George Parker, published in 1688. Chippendale used the process for some of his furniture but the fashion faded and only coach builders, and makers of inn signs and a few others, were left using the technique.

Turning to the question, it is, in my opinion, rather wasteful in time and money to make one's own lacquer. Like so many traditional products today, it is better to purchase them ready-made from trade suppliers. May I suggest that you use a good quality clear varnish or 'lacquer', such as japan varnish or polyurethane lacquer; these are fast drying varnishes and will do the job that you require. It should be pointed out that when I talk of lacquer in this context, it should not be confused with modern synthetic lacquers, which are totally different.

Workshop procedure:

1. Prepare the substrate to a fine smooth surface, i.e. damp down to raise the grain; fill; stop all imperfections, and sand down.

2. Apply one coating of spirit black pigment and allow to dry out.

3. Apply one coat of your choice of lacquer and allow to dry out.

4. Flat down this film using either 320 grade silicon carbide papers used dry, or 400 grade silicon carbide wet or dry used with soft water (rain or distilled) or white spirit.

5. Clean off with a dry cloth and apply further coatings of the lacquer, flatting down between each film of lacquer until you obtain a satisfactory surface. I would suggest 5–7 coats at least. Each coat must be allowed to harden off completely before the next is applied.

6. The final coating can be flatted down using 600 grade wet or dry papers, dusting off and burnishing with medium and fine burnishing pastes until a full mirror gloss finish has been achieved. It is best to use well washed mutton cloth for a burnishing cloth.

It should really go without saying that you should use only the best quality materials and good quality tools to achieve a first class finish. Make sure that you use the appropriate abrasive papers in the proper manner – the use of glass papers is not recommended. And last, but certainly not least, only top quality workmanship will produce good japanning.

Question

I have a range of new kitchen cabinets and shelves made of Douglas fir and I want to give them a stripped pine effect. Furthermore, I want a finish that can be easily cleaned. What do I do to achieve these requirements?

Answer

You say that the kitchen cabinets and shelves are new and made of Douglas fir, but you do not say if they are finished or just sanded ready for finishing. If the Douglas fir has been sanded, you must be on the look out for resin blisters; these exude sticky, yellow resin, which must be cleaned out using turpentine, and then sealed with shellac sealer or you will have trouble with the surface coatings.

To achieve a stripped pine effect to kitchen furniture is all very well, but you must look at it from the hygiene point of view. Kitchens are the workshops of the household, and steam, water, condensation, fat and grease build up on the walls, ceilings and units, so the finish applied must be one that cleans easily. A wax finish, for example, may be aesthetically pleasing, but is not very practical for the kitchen.

First, your wood needs to be stained to give it the appearance of stripped pine, which is darker than new pine. Make up a spirit stain using shellac sealer as a binder, mixed with methylated spirits (alcohol), together with a small amount of brown umber, and touches of yellow ochre and titanium white. Always try the colour first on offcuts of wood. Apply the stain by either spray gun, fad or mop.

Once you have achieved the effect you require, apply a coating of a pre-catalysed sanding sealer by spray, followed by a coating of pre-catalysed melamine lacquer, gloss or semi-gloss. This finish is ideal for kitchens and can be wiped down with soapy water. In my opinion, it is far better to use the semi-gloss effect on kitchen units rather than full gloss. If you do not wish to use a spray gun, you can obtain a 'brush-on' lacquer or use cellulose aerosols for the final finish.

Question

I have recently made a large hutch for my rabbits, and I am thinking of using yacht varnish for the bottom to prevent the urine sinking through. After reading an article about the risk of wood finishings, I should be grateful if you would advise me if yacht varnish will be safe for the job, especially as I want to breed rabbits. If not, what would you recommend?

Answer

What a charming question to answer!
   In considering the best advice to give you, I first consulted my local veterinary surgeon, and what follows is a combined effort, with the emphasis more on the health of the rabbits than on the protection of the wood.
   I am pleased that you suggest a yacht-type varnish, but even so, I do not recommend using any surface finish on a rabbit hutch, whether on the floor or anywhere else, simply because rabbits have sharp teeth and claws and would soon chew or scratch into the surface coating, rendering the water-proofing useless.
   It is suggested that a non-ferrous metal floor be fitted instead – something like a bird cage bottom, with a lip like a shallow tray, so that it can be removed from cleaning. This would be beneficial to the health of your rabbits and would remove the risk of them becoming addicted 'varnish sniffers'!

Question

I have some teak left over from a job and I am wondering if I could make a cutting board for the kitchen from the offcuts?

Answer

Teak is a wonderful wood. It is dull, odourous, oily, heavy and hard. It is resistant against wood eating insects, fire and acids, works well with hand and machine tools, has a lovely grain and figure, and can be used for a variety of purposes, such as shipbuilding, exterior and interior joinery, railway carriage work, furniture, etc. However, it is *not* suitable for kitchen cutting boards as it would be difficult to keep clean – bleaching would have little effect, so it would become greasy and un-hygienic for cutting foodstuffs very quickly.

Sycamore is an ideal wood for a kitchen cutting board, but it must not be finished with any surface coating. It will bleach and scrub white, however, so will be hygienic for use in the kitchen.

Question

Fourteen years ago our Parish church was treated for woodworm by professional workmen. At the same time, the church members obtained some of the chemical from the contractors and brushed it on to the pine pews on top of several coats of varnish. When the chemical dried, a coat of traditional varnish was applied. Unfortunately, the varnish did not harden completely and in hot conditions or long sermons people stick to the seat or back of the pew. Can you recommend an inexpensive cure for this problem?

Answer

I regret to inform you that there is no quick and easy way out of this unfortunately situation.

What has happened is that the chemical fluid has reacted with the layers of varnish beneath it as well as the coat of traditional oil resin varnish applied above it, leaving a sticky surface coating which will never really dry. The answer to your problem is to strip off all the varnish on your pews down to the bare substrate, in this case pine, with a chemical varnish stripper of the water washable type. I accept that this is no easy job, and I would recommend that you employ a local polisher for this work. When re-finishing is carried out, may I suggest that you use a shellac sanding sealer followed by a shellac varnish, which is spirit based and will not lead to the same problems that oil varnish can on pine or pitch pine.

However, there is one very easy cure to your problem — that is to ask the vicar to shorten his sermons!

Question

I am currently constructing some cabinets in oak. Unfortunately, the grain of the oak keeps lifting when I plane it. I have tried alternating the direction of planing, but the grain still lifts, leaving craters on the surface of the wood. Can you recommend a good wood filler that can be sanded down easily and will also take a stain, as I wish to finish the units in dark oak? Help, please!

Answer

Oak is not one of the easiest woods to work on. It requires very sharp tools to work the wood and I hazard a guess the reason why you are encountering such difficulties with the grain is that your tools are simply not sharp enough. In your particular case, I would recommend that you use a cabinet scraper, working in the direction of the grain, to remove all tool markings after planing. This will prevent lifting the top skin of the oak, which is causing the craters. The best type of wood grain filler, in my opinion, is a thixotropic type, which is easy to apply with old hessian. Work across the grain of the wood and allow it to dry. This type of wood filler is compatible with any type of staining product, including finishing surface coatings. I suggest that you also use aluminium oxide abrasive papers when working on oak as they have more bite than glass or garnet abrasive papers.

Thixotropic wood filler can be obtained in several colours from various trade houses, hardware stores, etc.

Question

I have been french polishing for some while now but I encounter difficulties when I reach the final finishing stage. I seem to get oily smears on the final surface, which increase as the polish dries. Where am I going wrong? I would appreciate your advice.

Answer

You do not say what oil you are using when french polishing; it could be either mineral oil or raw linseed oil. Your trouble stems from the fact that you are using far too much oil on your rubber. What is happening is that the oil is becoming part of the shellac film, which must be avoided at all costs. The purpose of the oil is to act as a lubricant for the rubber, and it should not be forced into the shellac film. The way to remove the smears is to use a little finishing spirit (not methylated spirits) in the spiriting-off stage, using your rubber as dry as possible, with straight strokes with the grain. This will, with a little practice, remove the surplus oil and restore the film to a gloss finish without smears. The skill is to carry on with the rubber until all traces of oil have been removed. What happens is that the spirit picks up the oil from the surface film, but there is a danger that if too much finishing spirit is used, it can 'ruck up' the polished surface, rendering it useless.

Another way of removing surplus oil is to use diluted sulphuric acid with vienna chalk in conjunction with a new rubber, but this is for experienced polishers only and I would recommend that you use the spiriting-off method.

Note: Finishing spirit is specially supplied and blended to be used to produce a perfect french polished, oil smearless, gloss finish.

Question

I have a limed oak dining suite which has done sterling service in our home for 29 years, and it has responded well to re-finishing. I achieved a satisfactory result by cleaning the surface and then painting it with white undercoat. However, two coats of paint were necessary to fill the grain adequately and the scraping required to remove the unwanted paint was very tedious. I am now making a wall cabinet to match the suite and have so far, even with the help of my local library, failed to discover a modern text which describes properly how a limed finish should be obtained. Please could you advise?

Answer

I am sorry to tell you but after reading your question it would appear that you have got the wrong idea about liming. The process involves colouring the grain of the wood, not filling it. The method was very popular in the 1930's, when hardwoods such as oak, ash and elm – that is, woods with large open grains – were coloured with a pigment, such as unslaked lime. A wire brush was rubbed into the grain of the substrate, to remove any soft tissue, followed by lime, which was allowed to dry off and then french polished over using white (which is really a clear finish) polish. Nowadays, 'lime' has been replaced by white liming paste polishes which can be purchased in tins ready for use. However, the wax polish must not be used with modern finishings; use instead a pigmented white paste which is compatible with lacquer spray surface coatings.

The process of 'liming' has been re-discovered in recent years, and all sorts of woods are now being limed, such as pine – in some cases to hide bad workmanship. If the process is carried out properly some very pleasing effects can be produced, as can be seen in many kitchen unit finishes. On oak or elm, if the background is coloured red, green or blue with a water stain and the surface then limed, the effect can be fascinating. Coloured 'lime', instead of white, can also be used to great effect.

The point I wish to make is that liming should allow the natural grain of the wood to show through the colouring effects, and must not on any account obliterate it.

Question

Three years ago we had a new conservatory built in iroko hardwood and this was finished with a clear varnish. However, the varnish has started to peel off and the whole conservatory looks very shabby on the outside. What can be done to restore the wood to its original condition, and what other, better type of finish do you suggest?

Answer

Iroko is a very hard wood which does not really require any type of finish. However, many of us are not happy unless we are coating wood with some kind of surface finish. In this case, I would not, under any circumstances, use an oil varnish. Iroko is a wood which will reject normal surface coatings, and with this in mind I suggest you use Danish oil. Other forms of oil will absorb dirt and grime, but this is not the case with Danish oil, which actually dries to a hard film. The procedure for restoring the wood in your conservatory is as follows:

1. Strip down to the bare wood using a chemical wood stripper. This is one of the safest ways of removing old varnish. Make sure that the surface of the wood is washed thoroughly with white spirit after the varnish has been removed.

2. When the surface is dry, sand down using 150 grade aluminium oxide abrasive sanding papers, and finish off with 240 grade garnet papers. Iroko does not in my opinion require any staining as the wood has a natural in-built colour.

3. If any filling has to be done, such as re-mastic filling around the glass, this is the time to do it. Any surplus filler should be wiped off with white spirit.

4. Apply the first coating of Danish oil with a soft, quality varnish brush and leave to dry hard for 24 hours.

5. When hard and dry, sand down using 320 grade wet and dry abrasive papers and a little water as lubricant. Wipe off residue and allow to dry.

6. Apply a second coat of Danish oil and leave to dry hard for 48 hours.

7. When hard, this finish can be further improved by using a little 0000 steel wool, thus flatting down to a smooth, denibbed finish.

This finish will not flake off like varnish and can be re-oiled at any time if it begins to look a little dry, simply by using a little of the Danish oil on a rag and wiping it over the surface. The whole job, incidentally, should be carried out during a dry spell for obvious reasons. If you follow this procedure, your conservatory will look good for many years to come no matter what the climatic conditions.

Danish oil contains tung oil and other blended synthetic resins, which makes it water resistant and ideal for exterior wood surface finishing.

Question

I am a builder and I have been asked to repair and restore a cedar shingle roof on a bungalow. The cedar tiles (shingles) are very dry and have a grey appearance. Some of the ridge shingles are covered with what looks like green slime. What is the best way of coping with this problem? I was thinking of treating the whole of the shingles with creosote as a finish.

Answer

First of all, you must not under any circumstances use creosote oil on cedar shingles as it will destroy the natural oils in the wood, and render its water resistant qualities useless.

To start with, you must replace the ridge tiles which are covered in slime as this will continue to be active. When replacing these tiles, make sure that you use galvanised nails as plain, untreated nails will rust within a short time when used with cedar. Once these repairs have been carried out, make sure that none of the other tiles are loose, after which you can proceed with the re-coating process.

As the cedar is very dry, it requires oiling. An average roof will take about 30–40 litres of clear thin oil, which can be obtained from some oil companies or the manufacturers of cedar built bungalows. Ideally, the oil should be applied with a 6 inch wide quality paint brush or applied by spray gun.

On no account should this process be carried out in damp conditions – it is best left for a very hot spell in the summer when the cedar will be bone dry. You will find that the cedar soaks up the oil very rapidly. Once the oil has been spread evenly over the roof area, it will take some time to soak in and dry out, but it will leave the shingles water resistant for the next 5–10 years. The whole roof will take on a lovely warm brown colour, although, with the passage of time, it will naturally turn grey.

The whole point of using oil is that you are replacing what has been washed out of the cedar over time.

Question

We are a company producing kitchen furniture and recently we have been experiencing difficulties in producing a quality finish using nitrocellulose semi gloss lacquer. We have a spray booth and all the necessary equipment, but the final finish seems to be cissing (fish eye). Could you give us some idea of where we are going wrong as this is taking up valuable time and we are wasting a lot of materials?

Answer

Your problem could be associated with a number of factors. As it has arisen at this time of year (winter), I think it could be caused by one of the following:

1. The temperature in the spray area may be too low. You say that you have a spray booth and it could be that the extraction is lowering the temperature. Therefore, you need to introduce warm, dry air into the spray area to increase the working temperature to around 70°F.

2. The spray booth could be drawing in air which is polluted with undesirable deposits, such as rubber, oil, silicone fumes, etc. Re-position the input air ducts to a cleaner location.

3. You may be using materials from various manufacturers and these may not be compatible with the lacquer viscosity requirements. Make sure that you use products from one supplier only.

4. Check your air lines from the compressor. These may be contaminated with water, which is produced in great quantity from piston compressors, and slight traces of oil. Make sure that you clean out the regulator and bleed the receiver at more frequent intervals.

5. You could add a percentage of anti-cissing agent to the viscosity, which will facilitate the production of a clean, wet film flow from the gun.

6. It could be that you are using too high a psi from the gun.

7. It could be that the stains you are using are not compatible with the lacquer. Oil stains are not recommended, so try a mixed solvent stain, or better still, use a water stain.

8. Certain woods react with lacquers in different ways. Try changing your nitrocellulose lacquer to a pre-catalysed semi-gloss lacquer which will in actual fact give you a tougher surface finish but it will behave like a cellulose finish for application.

9. Make sure that the sanding sealer is compatible with the top coating lacquer, i.e. it is from the same manufacturer.

10. Last of all, clean out all your equipment, including the spray booth walls. Cleanliness is of paramount importance if you wish to obtain a perfect finish.

If all the above fails, I would suggest that only one course of action is left open to you – sack the spray operator!

Question

At the moment, we are in the process of restoring our 18th century house, which has a room in it that is completely panelled in what looks like pine. The surface polish is very dirty and slightly water stained, but we do not want to strip off the original patina polish if at all possible. Do you think that we should strip off the polish or restore the surface? How do we tackle the project?

Answer

On no account would I strip the patina polish effect from your pine panelling. I suggest you follow the procedure outlined below:

1. Make up the following solution:
   Equal portions by volume of
   Linseed oil
   Paraffin oil (kerosene)
   White vinegar
   Soft water (distilled)
   Methylated spirits (alcohol)

Shake the solution well and apply to the surface with waste cotton wool. You will notice that the dirt and grime is easily removed. Continue until the cotton wool shows no further deposits of dirt.

This fluid will not harm the original polish in any way; it simply removes the dirt and leaves the original surface untouched and clean.

There may be areas of the panelling which are extremely dirty, in which case you will need to make up a further solution. This contains the same ingredients as before but a 5% amount of fine pumice powder is added to the mixture. This solution will remove stubborn areas of grime, but care should be taken not to cut into the original polish when using it. Alternatively, you could use fine steel wool (0000) with the original fluid formula on these areas.

The formula I have given is a standard antique reviver. If you do not wish to make it up, it can be obtained from various paint suppliers.

2. Allow the surface to dry out, which will take about 3–4 hours, and then apply a good quality wax polish with a mutton cloth. This will return the gloss finish polish to the now clean pine panelling.

There are some very good wax polishes on the market today and it is difficult to recommend one. Do not, however, use creams or spray-on waxes as they can cause difficulties in finishing in some cases.

Question

All our furniture has been in storage for over 15 months and now that it has been returned to us I am rather concerned about its condition. Most of it has a musty odour, and I am wondering how this can be removed. Furthermore, I note that the polished furniture has what looks like a bloom effect on it and I would welcome your advice on how this could be removed.

Answer

This is quite common when furniture has been stored for some considerable time, and what I suggest is that you give the whole of your furniture a spring clean. This would entail cleaning out every piece of furniture before you use it again.

First of all, wipe out the insides of drawers, cupboards, etc, with an anti-woodworm furniture polish. This will achieve two things; firstly, it will freshen up your furniture, and secondly, it will act as a woodworm deterrent. The backs and undersides of your furniture are just as important as the polished areas, so make sure that they are cleaned as well. If they show signs of green mould, then simply wash off using warm water and a little vinegar or mild ammonia. When dry use the polish to finish off.

As far as the bloom effect is concerned, this is common on highly polished furniture. It can be removed by either using furniture cream, or wiping the areas affected with a mixture of warm water and a little vinegar, followed by a good quality furniture polish when dry.

Your soft furnishings should be thoroughly cleaned with a vacuum cleaner, after which you should spray the undersides of chairs, settees, etc, with an anti-moth aerosol.

If, on the other hand, your furniture has been returned during the summer, then allow as much fresh air into the house as possible. Open all cupboards and drawers and allow fresh air to restore your furniture to its pre-move condition.

If you follow this procedure, your furniture should be restored to its original condition.

CHAPTER NINETEEN

Suppliers

UK Suppliers

Names and Addresses of Manufacturers & Suppliers	Outline of Products Manufactured or Supplied
R. Aaronson (Veneers) Ltd, 45 Redchurch Street, London E2	Veneer merchants
Fred Aldous Ltd, PO Box 135, 37 Lever Street, Manchester M60 1UX	Suppliers of craft materials, chair caning, leatherwork, etc.
Airflow Products Finishing Ltd, Northern Works, Underhill Lane, Sheffield, Yorks	Manufacturers of waterwash and dry filter spray booths
Axminster Power Tool Centre, Chard Street, Axminster, Devon EX13 5DZ	Woodworkers equipment and finishes
John Boddy Fine Tool & Woodwork Store, Riverside Sawmills, Boroughbridge, North Yorkshire YO5 9LJ	Specialist suppliers of timbers, finishes and woodworkers equipment
Chapman & Smith Ltd, Safir Works, East Hoathly, Lewes, East Sussex BN8 6EW	Manufacturers of protective equipment; gas & vapour respirators, etc.
Clam-Brummer Ltd, Maxwell Road, Borehamwood, Herts	Manufacturers of wood fillers
Craft Supplies Ltd, The Mill, Millers Dale, nr Buxton, Derbyshire SK17 8SN	Suppliers of tools, woods, equipment and finishes for woodturners
Cuprinol Ltd, Adderwell, Frome, Somerset BA11 1NL	Manufacturer of external wood preservatives, etc.
Cyclone Stripper Ltd, Phoenix Works, Avery Hill Road, New Eltham, London SE9 2BD	Manufacturer of 'Cyclone' paint stripper

DeVilbiss Ransburg Company Ltd, Ringwood Road, Bournemouth, Dorset BH11 9LH	Manufacturer of spray gun equipment, electrostatic guns, Duo-Tech system, etc.
Dorn Antiques, Tew Lane, Wootton, Woodstock, Oxfordshire OX7 1HA	Suppliers of leather skin, tooled for desk top lining
English Abrasives & Chemicals Ltd, Marsh Lane, London N17 0XA	Manufacturers of abrasive coated sheeting, belts, etc.
Fiddes & Son Ltd, Florence Works, Brindley Road, Penarth Road, Cardiff, South Glamorgan	General supplier of french polishes, lacquers, etc.
Gibbs Finishing Systems, Gibbs Road, Edmonton, London N18 3PB	Manufacturers and suppliers of specialist spray equipment – curtain coaters
Granyte Woodfinishers Ltd, Elton Street, Lower Broughton, Salford, M7 9TL	Manufacturers of industrial surface coatings including water-borne lacquers
Hydrovane Compressor Co Ltd, Claybrook Drive, Washford Industrial Estate, Redditch, Worcs B98 0DS	Manufacturer of rotor compressors
ICI Paints, Wexham Road, Slough, SL2 5DS	Manufacturers of paints, varnishes, etc.
International Paint, Powder coatings Division, Stoneygate Lane, Felling-on-Tyne, Gateshead, Tyne & Wear NE10 0JY	Manufacturers of powder coating materials
W.S. Jenkins & Co Ltd, Jeco Works, Tariff Road, London N17 0EN	Manufacturers of complete range of modern surface coatings
Kremlin Spray Painting Equipment Ltd, 839 Yeovil Road, Slough, Berks SL1 4JA	Manufacturers of Airmix, airless, electrostatic, robot spray guns, etc.
John Lawrence & Co Ltd, Granville Street, Dover, Kent CT16 2LF	Manufacturer of brassware used in the restoration trades
Liberon Waxes Ltd, Mountfield Industrial Estate, Learoyd Road, New Romney, Kent TN28 8XU	Suppliers of a complete range of finishing materials for woodwork and DIY
Magnum Compressors Ltd, Church View, Stockton-on-Forest, York YO3 9UP	Manufacturers of compressors and spray guns
John Myland Ltd, 80 Norwood High Street, West Norwood, London SE27 9NW	Suppliers of french polishes and modern surface coating lacquers, sundries
P & J Dust Extraction Ltd, Lordswood Industrial Estate, Chatham, Kent ME5 8PF	Manufacturers and suppliers of dust extraction equipment
E. Parson & Sons Ltd, Blackfriars Road, Nailsea, Bristol BS19 2BU	Manufacturers of varnishes, stains, french polishes, etc.
Permalite Ltd, Vincent Lane, Dorking, Surrey	Manufacturers of 'Jenolite' rust remover

E. Ploton (Sundries) Ltd, 273 Archway Road, London N6 5AA	Suppliers of brushes for the wood graining crafts
'Practical Woodworking' Magazine, King's Reach Tower, Stamford Street, London SE1 9LS	Monthly magazine
Rentokil Ltd, Felcourt, East Grinstead, West Sussex RH19 2JY	Manufacturers of preservatives and chemical insecticides for woodworm and rot
Romo Fabrics, Lowermoor Road, Kirkby in Ashfield, Nottingham NG17 7DE	Suppliers of white cotton felt and upholstery materials
Russell & Chapple Ltd, 23 Monmouth Street, London WC2H 9DE	Suppliers of muslin, artist brushes, canvas, paints, varnishes, etc.
Rustin's Ltd, Waterloo Road, Cricklewood, London NW2 7TX	Manufacturer of speciality paints, dyes, oils, paint remover, etc.
Sadolin (UK) Ltd, Sadolin House, Meadow Lane, St Ives, Cambs PE17 4UY	Manufacturers of interior and exterior surface coatings
Sonneborn & Rieck Ltd, Jaxa Works, 91–95 Peregrime Road, Hainault, Ilford, Essex IG6 3XH	Manufacturers of complete range of lacquers, powder coatings, acrylics, etc.
Sterling Roncraft, Chapeltown, Sheffield, Yorks S30 4YP	Manufacturers of varnishes, stains, waxes, etc.
Wilcot (Decorative products) Ltd, Alexandra Park, Bristol BS16 2BQ	Manufacturer of 'Nitromors' paint strippers
'Woodworker' Magazine, Argus House, Boundary Way, Hemel Hempstead, Herts HP2 7ST	Monthly magazine

North American Suppliers

Stains and Dyes, etc

Wood Finishing Supply Co, 100 Throop St, Palmyra, N.Y. 14522; (315) 597–33743.
Woodcraft Supply Corp, 41 Atlantic Ave, Woburn, MA 01888; (800) 225–1153.
The Woodworkers Store, 21801 Industrial Blvd, Rogers, MN 55374–9514; (612) 428–2199.

Varnishes, etc

McCloskey Co, 7600 State Rd, Philadelphia, PA 19136
Fuller–O'Brien, 450 E Grand Ave, S, San Francisco, CA 94087.
Pratt & Lambert Inc, Box 22, Buffalo, NY 14240.
Benjamin Moore & Co, Montvale, NJ. (Moore's and Benwood are trade marks of this company)
(These products are obtainable from most hardware stores.)

Spray Guns and pneumatic equipment

Apollo Sprayers, 1030 Joshua Way, Vista, CA 92083
Binks Manufacturing Co, 9201 W Belmont Ave, Franklin Park, IL 60131.
The DeVilbiss Co, Box 913, Toledo, OH 43692.
W.W. Grainger Inc, 5959 W Howard St, Chicago, IL 60648.
Grumbacher, M. Inc, 30 Engelhard Drive, Cranbury, NJ 08512.
Amspray (American Spray Industries) 221 South State Street, PO Box 86, Harrison, OH 45030.
Wagner Consumer Products Division, 1770 Fernbrook Lane, Minneapolis, MN 55447.

Safety Equipment

Reliable Finishing Products Inc, 2625 Greenleaf Ave, Elk Grove, IL 60007.
Compliant Finishing Equipment, Graco, Inc, PO Box 1441, Minneapolis, MN 55440–9276.

Cream Waxes, etc

Roger A. Reed Inc, Box 508, Reading, MA 01867.

Lacquers, sealers, thinners, etc

Grand Rapids Woodfinishing Co, 61 Granville, Ave, S.W., Grand Rapids, MI 40503.
Randolph Products Co, Park Place East, Carlstadt, NJ 07072.
H. Behlen & Bros, Route 30 North, Amsterdam, NY 12010.
Lee Valley Tools Ltd, PO 6295 Station J, Ottawa, Ontario K2A 1T4.

Index

abrasive papers:
  aluminium oxide, 13, 19, 20, 21, 23, 24, 48, 90
  emery, 19, 22, 23
  garnet, 13, 17, 19, 20, 23, 24, 42, 48, 56, 61, 90, 101, 102
  glass, 19, 20, 23, 24
  lubricated silicon carbide, 19, 21, 23, 24, 25, 136, 143, 144, 152, 181
  Lubrisil, 21, 60, 90, 91, 92, 106, 130, 142, 143, 179
  tungsten carbide, 22
  wet and dry, 19, 21, 24, 67, 69, 87, 137, 144, 181, 182
acetic acid, see acids
acids:
  acetic, 32, 36, 39, 52, 58, 82, 83, 180
  hydrochloric, 145
  hydrogen peroxide, 39, 40
  nitric, 82, 83, 122, 135
  oxalic, 41
  phosphoric, 66, 70
  pyrogallic, 82, 84
  sulphuric, 82, 83, 104, 122, 135, 145
  tannic, 82, 87, 112, 133
acid catalysed lacquer, see lacquers
acid catalysed shellac garnet polish, 97, 98
acid catalyst, 146
acid finish, 102, 104
acrylic lacquers, see lacquers
adhesives, 15, 42, 49, 78, 169
aerosol plastic cleaners, 32, 147
aerosol wax, 64
air cap, 161
air compressors, 164, 165, 166, 169
air condenser, 166
air hose, 162, 169
air regulator, 166, 169
air transformer, 156, 166
airless system, 51, 156, 168, 169

airmix system, 156, 169, 170
alcohol, 81, 136
alkalis, 39
aluminium oxide papers, see abrasive papers
ammonia, 39, 40, 51, 60, 61, 78, 82, 83, 84, 87, 131, 133
angelo brothers, 96, 97
aniline dyes, 78, 96, 131
anti-cissing fluid, 128, 180
antiquing, 87
antique paint effect, 123
antique wax polish, 63
antique worn pine effect, 123
ash, 39, 41, 91, 116
automatic spray system, 195

beaumontage wax, 14, 61
beech, 39, 41, 119
beeswax, 60, 61, 62, 63, 92
beezer, 23, 102, 103
bench sander, 25
bichromate of potash, 82, 84, 112, 122, 133
bleached montan wax, 62
bleaching, 38, 39
bleeder-type gun, 160
blooming, 176, 180
blushing, 180
bodying in, 101, 102, 104
boiled oil, see oils
brass wire brush, 69
brood lac, 97
bruises
brown umber, 61, 99, 109, 111, 112, 115, 123
brushes:
  brass wire, 69
  dragger, 124
  fad, 79, 81, 94
  fan pencil overgrainer, 124
  flogger, 124
  grainer, 123

Index

grass, 48
liner, 121
mop, 81, 99, 121
mottler, 121, 124
softener, 123
stipple, 121, 123
sword liner, 124
brush-on french polish, 106, 107
brush-on lacquer, 138
burnishing, 70, 92, 137, 143, 146, 150, 153
burnt sienna, 74
butyl acetate, 180
butyl alcohol, 180
button polish, 98

cabinet scraper, 14, 49, 100, 101
camphor oil, see oils
candelilla wax, 62
carbon effect, 125
carbon tetrachloride, 66, 69
carnauba wax, 61, 62, 63, 92
castor oil, see oils
caustic soda, 39, 42, 50, 82, 84, 130
cedar, 32, 41, 56, 58, 128, 132
cellulose lacquer, see lacquers
cellulose thinners, 48, 69, 128
ceresin wax, 62
chemicals (glossary), 82
chemical stains, 81, 94, 115
cherry, 41
chestnut, 41, 82, 84, 87, 116, 122
chilling, 181
Chinese wood oil, see oils
Chinese wax, 62
chipboard, 17
cissing, 128, 132, 155, 158, 179
cleaning
cleaning fluids, 36
coal tar stain, 88
coating faults, 179
coconut, 32
cold galvanizing, 67
collage, 126
colour matching, 74
coloured french polish, 98
combination stains, 85
common furniture beetle, 31, 35
compressed air, 155
copper sulphate, 82, 83
corrosion, 68
COSSH regulations, 25, 196, 197, 198, 199

crackle glaze, 127
creosote oil, 37, 56, 58, 89
cross linker, 152, 153
curtain coater, 141, 192, 193, 194

Danish oil, see oils
death-watch beetle, 31, 35
dermatitis:
allergic contact, 201, 202
direct irritant contact, 201
DeVilbiss Ransburg Duo Tech system, 156, 173, 176
Devilmix, 171
diaphragm pump, 168
dipping tank, 91
distressing, 23, 109, 116
dragger brush, 124
dry-back spray booth, 189
dry rot, 35, 36
dry spray, 179
dust extraction, 94, 205
dyes, 76, 77
dye, aniline, 78, 96, 131

ebony, 32, 41, 98
edge mop, 150
electrostatic spray system, 71, 156, 170, 178
elm, 41, 91, 116
emery paper, see abrasive papers
emulsion wax polishes, 19, 64
Environmental Protection Agency, 196
etch primer, 68
eucalyptus oil, see oils
extraction systems, 189
eyes, first aid, 202

fad, 79, 81, 94
fan pencil overgraining brush, 124
Far Eastern effect, 124
fat edges, 181
ferrous sulphate, 82, 83
faults, 179, 180, 181, 182
fillers, 13, 14, 15, 16, 17, 34, 69, 100, 101, 116, 133
F.I.R.A., 153
fire extinguishers, 203, 204
fire fighting equipment, 203, 204
first aid, 202, 203
flash point, 136, 148
flexible sander, 25
flock spray effect, 125

flogger brush, 124
flooding, 182
fluid adjustment valve, 162
fluid needle, 161, 163
fluid tip, 161, 162
fluidised bed, 71
fog effect, 112, 125, 126
ford cup, 139, 146, 152, 159
frass, 31
french chalk, 23, 180
french polish, 16, 17, 25, 46, 49, 55, 57, 59, 60, 90, 96, 100, 102, 106, 112, 119, 123, 130, 131, 132, 133
french polish rubber, 119
fuming, 83, 84
fumigation tent, 32, 84
fuss, 81, 82, 94, 116

garnet paper, see abrasive papers
garnet polish, 97, 98
glass paper, see abrasive papers
glaze finishes, 115
gold leaf, 73
grainer brush, 123
graining comb, 124
graining effect, 123
grass brush, 48
gravity feed gun, 160, 168

health and safety, 17–18, 29–30, 37, 44–45, 54, 64–65, 72, 89, 94–95, 108, 127, 154, 184, 204, 206
hessian, 15, 115, 121
high-build friction polish, 91, 92, 94
hot air gun, 50
hot spray application, 169, 176
hydrochloric acid, see acids
hydrogen peroxide, see acids
Hydrovane, 168

ingestion of chemicals, 202
ingestion of poisons, 202
inhalation of fumes, 202
iroko, 32, 39, 56, 128, 131
isocyanates, 147, 148, 149

Japan wax, 61
jarrah, 131

knotting, 99, 129
knots, 17

Kremlin, 156

lac, 96
lac wax, 62, 98
Laccifer lacca, 97
lacquers:
  acid catalysed, 15, 16, 25, 34, 42, 46, 59, 77, 78, 115, 119, 122, 126, 130, 138, 141, 142, 145, 146, 147, 148, 151, 154, 158
  acrylic, 25, 78, 86, 87, 115, 129, 131, 151, 152, 153, 154
  cellulose, 25, 34, 42, 46, 47, 55, 59, 60, 61, 67, 69, 70, 77, 78, 79, 81, 90, 91, 100, 112, 115, 126, 136, 137, 138, 140, 145, 151, 154, 180
  nitro cellulose, 57, 71, 135, 136, 141
  polyester, 16, 22, 25, 46, 49, 149, 150, 154
  polyurethane, 16, 46, 47, 59, 147, 148, 154
  pre-catalysed, 16, 17, 22, 25, 34, 59, 60, 78, 100, 115, 119, 122, 126, 129, 130, 141, 142, 143, 145, 151, 154, 158
  waterborne, 25, 55, 151, 172
lacquer patina finish, 143
lamp black, 74, 111
leather, 112
lime, 91, 98
lime finish, 116
liming, 73, 119, 133
liming wax, 63
liner brush, 121
linseed oil, see oils
longhorn beetle, 31, 35
lubricated silicon carbide papers, see abrasive papers
Lubrisil, see abrasive papers
lye, 39, 42, 50, 82, 84, 130

mahogany, 39, 41, 73, 74, 82, 84, 91, 98, 119, 129
maple, 41
marbled finishes, 121, 186
masking effects, 124
MDF, 13, 17, 121, 134, 141
melamine, 141, 145
methylated spirits, 46, 49, 51, 79, 98, 99, 100, 102, 104, 180
methylene chloride, 47
mineral oils, see oils
mixed solvent stains, 38, 81, 137, 139
mop, 81, 99, 121
mottler brush, 121, 124

Index

multi-colour effects, 183
mutton cloth, 63, 87, 105, 123

naphtha stains, 80, 81
netting, 183
nibs, 181, 182
nitric acid, see acids
nitro-cellulose lacquer, see lacquers
non-bleeder spray gun, 160
non grain-raising stains, 81
non-reversible finishes, 47

oak, 39, 55, 56, 73, 82, 83, 84, 87, 91, 116, 122, 128, 133
oils:
  boiled, 55, 57, 131
  camphor, 58, 61
  castor, 56, 57
  Chinese wood oil, 56, 57
  creosote, 36, 57, 58, 89
  Danish, 56, 58, 92, 93, 129, 131
  eucalyptus, 32, 41, 51, 135
  linseed, 51, 55, 57, 59, 131
  mineral, 57, 58, 129, 131, 132
  paraffin, 36, 51, 58, 63
  poppy, 57, 59
  rape seed, 57, 59
  raw linseed, 52, 56, 57, 59, 92, 99, 180
  Scandinavian, 56, 92, 93
  teak, 56, 58, 92
  tung, 56, 58
  turpentine, 58, 60, 63
oil stains, 38, 77, 78, 81, 115
orange peel, 158, 181
oxalic acid, see acids
ozokerite wax, 62

paint, 25, 50, 69, 121
pale polish, 98
paraffin oil, see oils
paraffin wax, 61, 64, 150
patina, 22, 51, 70, 92, 111, 112, 115, 144
permanganate of potash, 82, 83
phosphoric acid, see acids
piano finish, 23
pickled pine, 83, 84, 121
pigments, general, 17, 73, 74, 75–76, 79, 80
pigments:
  brown umber, 61, 99, 109, 111, 112, 115, 123
  burnt sienna, 74
  lamp black, 74, 111
  red lead, 67, 132
  titanium white, 74, 123
  venetian red, 74
  yellow ochre, 74, 112, 123
pine, 39, 41, 50, 56, 77, 98, 109, 119, 122, 129, 130, 135
piston pumps, 168
piston type compressor, 164
pole lathes, 90
polishes:
  a.c. shellac garnet, 97, 98
  antique wax, 63
  beeswax, 60
  button, 98
  coloured french, 98
  high-build friction, 91, 92, 94
  liming wax, 63
  pale, 98
  white french polish, 98
polyester lacquer, see lacquers
polyester powder, 71
polyurethane lacquers, see lacquers
poplar, 41
poppy oil, see oils
pounce bag, 103
powder coating, 71
powder post beetle, 31, 35
powered respirators, 95
pre-catalysed lacquer, see lacquers
preserving stains, 88
pressure feed gun, 168
pressure feed system, 163
primers, 147
printing, 182
pullover chemical solvent, 135, 137, 144, 145
pullover fluid, 144
pullover rubber, 135, 137, 141, 142, 144, 153, 181
pumice powder, 23, 52, 103, 112
putty, 14, 15
pyrogallic acid, see acids

rag rolling, 123
rape seed oil, see oils
raw linseed oil, see oils
red lead, 67, 132
redwood, 32
remote pressure feed cup, 163
Rentokil, 33
retarder thinners, 180, 181

reviver, 51, 58
robots, 156, 195
rolled edges, 181
roller coating machine, 141, 194–195
rosewood, 41, 78, 98, 132
rotary compressor, 164
rotor compressor, 168
rottenstone, 23, 52, 64
rubber, 101
rucking, 182

safety, 199, 200
   (see also health)
sand blast effect, 120
sander, computer, 28
sander, disc, 28
sander, orbital, 28
sander, universal, 28
sander, wide panel, 28
sanding block, 14
sanding machines, 25, 50
sanding sealers, 94, 97, 100, 111, 122, 126,
   128, 133, 138, 139, 140, 141, 146, 152, 172
satinwood, 41
Scandinavian oil, see oils
scrubbed pine effect, 122
scumble effect, 123
seed lac, 97
sfumato effect, 125, 126, 183
shaft flexible sander, 25
sheens, 147
shellac, 22, 52, 77, 79, 80, 96, 97, 98, 99, 104,
   111, 130
shellac cut, 98
shellac finish, 32, 40
shellac sealers, 13, 77, 94, 98, 106, 115, 130
shellac stick, 61
shot blasting gun, 51, 66
silicon wax, 62, 63
skin burns, 202
skinning in, 101
slaked lime, 84
softener brush, 123
soldering iron, 17
solvents, 59, 66, 77, 81, 179
spattering, 112, 182, 183
splatter effect, 120
spiriting off, 101, 102
spirit stains, 38
spray booths, 71, 147, 158, 179, 188
spray finishing, 178

spray guns, 71, 79, 81, 99, 106, 112, 115, 119,
   120, 121, 126, 133, 155, 159
spray stains, 88
spray tables, 191, 192
spraying techniques, 141, 177
spreader adjustment, 162
spruce, 135
stains, general, 76, 77, 80
stains:
   bichromate of potash, 82, 84, 112, 122, 133
   chemical, 81, 94, 115
   coal tar, 88
   combination, 85
   copper sulphate, 82, 83
   ferrous sulphate, 82, 83
   mixed solvent, 38, 81, 137, 139
   naphtha, 80, 81
   non grain-raising, 81
   oil, 38, 77, 78, 81, 115
   permanganate of potash, 82, 83
   preserving, 88
   spirit, 38
   spray, 88
   varnish, 85
   water, 38, 78, 115
stainers, 74, 120, 121
Statech gun, 170
steel wool, 22, 46, 48, 56, 70, 87, 105, 107,
   112, 115, 123, 143
stencilling effect, 120
stick lac, 97
stipple brush, 121, 123
stoppers, 14
stoving oven, 71
stripping compounds, 48, 49, 52
substrate, 13, 38, 46, 48, 110, 116, 119, 137,
   139, 141, 155, 161, 170, 177
suction feed gun, 156, 159, 160, 168
suede effect, 184
sulphuric acid, see acids
sword liner brush, 124
sycamore, 39, 41

tak rag, 181
tannic acid, see acids
teak, 32, 38, 56, 58, 78, 128
teak oil, see soil
thermoplastic, 71
thermosetting powder surface coating, 71
thinners, 59, 138, 142, 146, 151, 180
titanium white, 74, 123

Index

touching up, 80
trigger, 162
trolley racks, 194
tung oil, see oils
tungsten carbide papers, 22
turpentine oil, see oils
two-pack bleach, 39

ultra violet, 92, 130, 150
urea formaldehyde, 141

varnishes, 19, 25, 50, 52, 55, 59, 78, 80, 86, 87, 92, 121, 133
varnish stains, 85
veneers, 42, 49, 76, 78, 116, 134, 141
Venetian red, 74
Vienna chalk, 23, 83, 104
viscosity, 136, 138, 142, 144, 146, 148, 151, 152, 155, 158, 159
V.O.C. emissions, 170, 196, 197

walnut, 73, 84, 91, 98, 129
washing soda, 82, 84
waterborne lacquers, see lacquers
water stains, 38, 78, 115
waterfall effect, 183, 184
wax stains, 87
waxes, general, 19, 51, 78
    aerosol, 64
    beaumontage, 14, 61
    beeswax, 60, 61, 62, 63, 92
    bleached montan, 62
    candelilla, 62
    carnauba, 61, 62, 63, 92
    ceresin, 62
    Chinese, 62
    emulsion, 19, 64
    Japan, 61
    lac, 62, 98
    ozokerite, 62
    paraffin, 61, 64, 150
    silicon, 62, 63
wadding, 100
waxing, 62
wet rot, 35, 36
Western red cedar, 39
wet and dry paper, see abrasive papers
wet back spray booth, 189
white french polish, 98
white spirit, 36, 52, 58, 60, 66, 67, 69, 180
woodturning, 90, 92
woodworm, 32, 36, 109, 110, 129
woodworm treatment, 32
woodworm fluid, 32

yellow ochre, 74, 112, 123
yew, 133

zebrawood, 41
zinc chromate metal primer, 67
zinc phosphate, 66, 68
zinc phosphate metal primer, 67
zinc stearate, 140

Books of Related Interest

Making Wooden Clock Cases *Tim & Peter Ashby* — Designs and plans for 20 clocks from the Ashby Design Workshops. Complete measured plans in metric and imperial measurements with full instructions for a wide variety of traditional clock cases. Includes a full range of suppliers for clock movements and other components.

World Woods in Colour *William Lincoln* — 275 commercial world timbers in full colour, describing general characteristics, properties and uses table. 300 pages.

Spindle Moulder Handbook *Eric Stephenson* — Covers all aspects of this essential woodworking machine from spindle speeds to grinding and profiling. 200 pages — 430 photos and line drawings. (**Shaper Handbook** in the US)

The Conversion & Seasoning of Wood *Wm. H. Brown* — A guide to principles and practice covering all aspects of timber conversion from the log and dealing with proven methods of seasoning. 222 pages illustrated.

The Marquetry Manual *Wm. A. Lincoln* — This state-of-the-art publication incorporates all the traditional ideas and practices for marquetarians as well as all the current thinking, and a selection of some of the greatest marquetry pictures. 272 pages, 400 illustrations.

Relief Woodcarving and Lettering *Ian Norbury* — Caters for all levels of ability from beginners onwards, exploring the fields of low and high relief carving through a series of graded projects. 157 pages, fully illustrated.

Modern Practical Joinery *George Ellis* — This vast coverage of internal joinery includes windows, doors, stairs, handrails, mouldings, shopfitting and showcase work, all clearly detailed and illustrated with hundreds of line drawings. Nearly 500 pages and 27 chapters.

Mouldings and Turned Woodwork of the 16th, 17th & 18th Centuries *T. Small & C. Woodbridge* — This large format book presents full size details and sections of staircases, doors, panelling, skirtings, windows, together with architectural turnings and many other specific applications of mouldings.

Books of Related Interest

Circular Work in Carpentry and Joinery *George Collings* — A practical guide to circular work of single and double curvature. 120 pages, fully illustrated throughout.

Modern Practical Stairbuilding and Handrailing *George Ellis* — Another epic work by Ellis devoted especially and seriously to the demanding art of stairbuilding and handrailing. 352 pages and 108 plates.

British Craftsmanship in Wood *Betty Norbury* — Nearly 200 designer/makers are represented showing exacting standards of work and a creativity unsurpassed in the art of woodworking. 192 pages. Illustrated with colour and black and white photographs. Hardcover. (**Fine Craftmanship in Wood** in the US).